Fool Willing

Advance Praise

"As one who has spent years building community with an environmental focus in diverse urban settings, it is refreshing to read *Fool Willing*. The book is smart, creative, and approachable for all walks of life and all members of the community no matter their background. As a leader in the Green Movement, I consider this book perfect for my needs. Kathy's powerful experience, derived from the 24/7 responsibility of raising a child with a disability, comes through in the form of patience, acceptance, and her understanding about judgment and prejudice from others. The practical exercises designed to get people into a playful attitude at work are brilliant. Using play as a way to cross the many divides between us is timely and important. I'm already making a list of people I know who need to read this book."

–**Ken Leinbach**, Executive Director, Urban Ecology
Center of Milwaukee, Urbanecologycenter.org

"What a pleasure to read! It's a bright twist on coaching and service work. I just love it! I had such a chuckle with the story about 'Grant.' I've done grant writing and it would have been a lot more fun to write to 'Grant'! I found myself surprised and enchanted at so many of the teachings in these pages, even as a certified Martha Beck coach who knows how much Martha uses play. This book is going to rock the house."

–**Shannon Conway**, Coach and Entrepreneur,
Belly-mind.com

"In *Fool Willing*, Kathy Oppegard provides us with an inspiring, practical, and joyful roadmap for enriching and broadening environmental action. In this delightful guidebook, she taps into a brilliant strategy for creating a sustainable environmental movement: a dual emphasis on the roots of human connection and human creativity. If you're seeking novel, fun approaches to growing and nourishing your green organization, this book will provide you with many practical methods for inviting newcomers and strengthening the bonds between volunteers, board members, and staff."

–**Kris McGuffie**, Writer, Editor and Writing Coach,
Metamorphosis, InspireMetamorphosis.com

Fool Willing

The Secret Power of Play
to Engage Communities in
Your Green Organization

Kathy Oppegard

NEW YORK

NASHVILLE • MELBOURNE • VANCOUVER

Fool Willing
The Secret Power of Play to Engage
Communities in Your Green Organization

Published in New York, New York, by Morgan James Publishing in partnership with Difference Press. Morgan James is a trademark of Morgan James, LLC.
www.MorganJamesPublishing.com

The Morgan James Speakers Group can bring authors to your live event. For more information or to book an event visit The Morgan James Speakers Group at www.TheMorganJamesSpeakersGroup.com.

Brand and product names are trademarks or registered trademarks of their respective owners.

ISBN 978-1-68350-500-6 paperback
ISBN 978-1-68350-500-6 eBook
Library of Congress Control Number: 2017904088

Cover Design by: Rachel Lopez, www.r2cdesign.com
Interior Design by: Bonnie Bushman, The Whole Caboodle Graphic Design
Editing: Grace Kerina
Author's photo courtesy of Lea Wolf
Photo of dog with Frisbee courtesy of Alexander Dummer, Unsplash
Photo of open book courtesy of Ben White, Photographer, Unsplash
Photo of woman in leather pants and girl courtesy of London Scout, Unsplash
Photo of clock courtesy of Katarzyna Kos, Unsplash
Photo of tree courtesy of Reza Shayestehpour, Unsplash
Photo of sunflowers courtesy of Ryan Waxberg, Unsplash
All other photos courtesy of Kathy Oppegard

In an effort to support local communities, raise awareness and funds, Morgan James Publishing donates a percentage of all book sales for the life of each book to Habitat for Humanity Peninsula and Greater Williamsburg.

Get involved today! Visit
www.MorganJamesBuilds.com

Dedication

I dedicate this book, with love, to David,
Finnigan, Bjorn, and Annika Rose, and to the
wonderful diversity of this beautiful planet.

Table of Contents

Introduction

The Secret Power to Building Community

Communities. It's such a simple word to represent something so intricate, so interwoven. There are communities of people and communities of trees and plants and animals. We can look at an ecosystem as a community.

If you're like me, you've often wondered how to create long-term community engagement in the environmental movement and with your organization. Maybe there are certain pockets of your membership or team that are involved, but others have not engaged at all, are not part of this process. And you wonder, *How do I get people to care about the planet? What can I, as one person, do to change our trajectory? Is there a way to make this process of creating community more fun? Can we change the world and save the planet—and have a grand time, too?* I say *yes*!

You may feel tired, and like you're already doing everything that you know to do. And yet, giving up is just not an option. We love this wonderful planet, and there are beloved people and animals that we share it with. We're on the lookout for signs of hope and laughter and progress.

Long-Lived Communities

We want to engage people over the long-term to care about the quality of the air they breathe, the water they drink, and the land they live on. We want our and their children to be healthy, playful, thriving, and connected to the world. We want to be and to raise excellent caretakers of the world.

So, how do we do this?

Sometimes, it seems like people are too involved with their devices, with social media, with connecting in every way except face-to-face or with the land that surrounds and supports them. We know that parts of our communities are not included in our efforts; that there are groups that are not at the "green table." We need more community engagement and more diverse communities throughout our environmental organizations.

What if play is the secret power you have to get them there?

A Community of Change Agents

It can seem like all we see in the media about the environment are doom and gloom scenarios. Yes, there are major problems— *truly* epic problems: climate change, a massive extinction rate, the pollution of air and water, the degradation of farmland, and environmental racism. (*I hear you. The temptation to get overwhelmed* is *overwhelming.*) But, also, there are good people everywhere working to solve these issues and more, *and having fun doing it.*

Those stories are not found on the front pages. For example, there is the local farm-to-table movement, which helps with food security, decreases "food miles" (how far your food travels to get to your plate) and so mitigates climate change, uplifts local economies, and improves nutrition. And there is the sheer fun of watching kids eating their first corn on the cob grown by a local farmer, or laughing with them as watermelon juice drips down their faces.

In so many ways, we, as environmentalists and community leaders, can help people move beyond the fear to paint a picture of environmental hope for a broader variety of people—and make that picture a reality.

Solving environmental problems is complex, and there are no silver bullets, no simple answers. But there are skills we can use, and there are impactful, meaningful things we can do to create and build thriving communities and environmental resiliency. We can bring our natural gifts as humans—as storytellers, inventors, healers, dancers, and poets—into the arena of problem-solving environmental challenges. We can *enjoy* the process and help others do the same. Our creativity, our delight, and our joy can build right along with our community's capacity for positive change.

That's what this book is about: supporting people in creating *their own* picture of hope, for themselves and for the natural world, and adding a motivating dose of play to turn that picture into reality.

Learning from Nature

Let's be real. We *are* in it for the long-term. We are here because the seventh generation means something to us, and we want to be proud of the legacy we are leaving. We want the water to be more clear, the land to be more healthy, the air to be more pure than it is now for the generations to come. We could talk endlessly about the problems, but let's focus on sharing solutions and developing microcosms that work and are replicable.

A way to do this is to learn from the natural world, to notice the patterns there, and to do our best to imitate them, for they have long functioned beautifully in solving problems.

We can also look to the times when people naturally gather in community—to relax and have fun, to enjoy themselves and one another—as a method for gathering ways to add joy and adventure to our endeavors.

Real Community

Does this sound something like you? "I'm not interested in being a knight on a white horse, coming in to 'save the day.' The knight wants to get out of that heavy armor that's too darn hot, and the horse wants to graze and rest. As a leader, I'd like my communities to understand that I respect their needs and I'm not interested in one-shot solutions. I want to engage in ways that are meaningful and long-term and, frankly, more fun; ways that create community—a real community of people who laugh and learn and argue together and support one another. I want people to look more toward each other for leadership and support and ideas."

We can remind the people around us that they don't have to have advanced degrees to make a real difference. They may need to have patience and persistence and the capacity to encourage and laugh at themselves and each other, but those strengths are part of our legacy as humans.

We can help by reminding our communities of that.

My Why for Writing This Book

My four-year-old daughter squawks with laughter at the sight gags her older brother pulls—a silly dance, a surprising face. All get a loud laugh from her. She delights in meeting new people and has big, sparkly, welcoming smiles for them. She loves people, and they tend to love her right back. She is smart and funny. If you make faces or are expressive with your gestures when you speak, she thinks that's hilarious.

When she was a tiny girl and I was holding her in my arms, sometimes she would lay her head on my shoulder. People would smile at us, and say, "Oh, somebody is sleepy." My throat would get kind of tight and I would nod. Inside, I was thinking, *Well, actually, she's really tired from holding up her head. She can't do it as long as a typically developing kid her age, because she has quadriplegic cerebral palsy.* At some point, I realized that my daughter was "passing" as a "normal" kid. That made me think about my friends who are people of color and who must wade daily in the stinky waters of racism, and how they do not have the opportunity to "pass."

My daughter uses a walker. It's obvious that she moves differently than other people. While walking with her, sometimes people glance at us, and then avoid looking at her on the sidewalk. It's shocking and hurtful to be looked at, then shunned by shuttered eyes and heads turned away. And yet… so many more people go out of their way to say hello to her, to smile at us, to welcome her. This is the good heart of humanity.

I am my daughter's advocate. I have had to be fierce and persistent and kind and educate people and check my own biases—and repeat this on a daily basis. I want this world to work for her, too. She is such a delightful little bean who gets jokes, and makes them, and brings such joy into the world for many people.

I'm dedicated to making the process of including her and diverse groups of people, making our world better, and making positive change more fun for everyone.

When I've asked myself, *Who am I to write such a book as this?,* the answer that comes back is, *Who am I not to?* If I can make the process of growing a diverse and connected community in the environmental movement easier and more fun and playful for you, then I will have helped not only my daughter, but others, too, and that is good.

This book is a song in honor of community; in honor of all the wondrous diversity of humanity; in honor of anyone who has felt the sting of being considered "the other" in a community.

I invite you to the "green table", with joy. You belong here. We all do.

Annika and me laughing in the grass at
Donald Park, near Mt. Horeb, Wisconsin

About Your Guide

I'll tell you a little about my involvement with environmental change and with using play as a tool for engagement and change.

As the coordinator for America Recycles Day and Pollution Prevention Week for the Department of Natural Resources of the State of Wisconsin, I helped coordinate the efforts of dozens of community organizers throughout the state, who then coordinated hundreds of volunteers. Some of our efforts included collecting over seven tons of clothes for Goodwill Industries, more than twenty tons of computers and equipment for re-use and recycling, and coordinating lake clean-ups, additional recycling efforts, and much more.

On the University of Wisconsin-La Crosse campus, right in the middle of campus, volunteers did a dumpster dive, and then sorted out recyclables from the garbage. They showcased the process, highlighted the room for improvement, and invited students to do better in their recycling. Their willingness to dive in and get messy, to play with the garbage and recyclables, got them good press and kudos.

In a small town in northern Wisconsin, people were thrilled to donate good, used clothes to Goodwill, and it became a big "help-your-neighbor clear their closet" effort. People got all giggly and elated. One participant told the organizer, "Oh, I've been saving these clothes from my grandparents for years. I'm so glad to give them a good home now." People asked their friends and neighbors if they had clothes to donate and, together, they collected over seven tons of clothes in a day! They brought their generous haul to a semi-truck in a parking lot. That was in northern Wisconsin on November 15, when it's cold! The community did something good, did it together, and had a great time doing it.

I have helped coordinate the efforts of hundreds of volunteers for Wisconsin's Folklife Festival, a celebration of the folkways of the multitudes of peoples of our state.

As the Volunteer Coordinator for Hammer with a Heart for Project Home, I was honored to coordinate over 1,000 volunteers who helped make needed major home repairs. We set a tone of respect when in someone's home, and invited people (both the helpers and the helpee) to talk, make jokes, and work together.

I have seen firsthand the generosity of the stranger, who touches the life of a person who needs help. I believe our hearts and minds and communities are stronger than the problems we face. Our capacity to laugh, connect, and problem-solve will bring us through.

Your Community Experts

I am not the expert on the communities in your area. *The people in those communities are the experts.* You can reach out to them, and this book will help you do that. I invite you to reach out to them in ways that are playful and appealing—and this book will give you ideas about how to do that.

Introducing play to your community outreach efforts is like sharing new dance moves. It is fun when people share their new moves, take turns, and then dance together in a big group. Dance becomes a way to play together, to laugh and connect with each other, and to applaud one another's creativity, expressiveness, and joy.

In this book, I share simple, fun, and effective ways to engage community and, in the process, invite more diversity to the table. When you start this kind of conversation and build a plan that includes both diverse communities and play, you create even more opportunities for growth, learning and creativity—*and the playful community spreads.* This builds a true and lasting positive change in our communities, to include and protect this beautiful planet we all share.

First let's find out how renewing your energy is the best first step to renewing the world.

Chapter 1
Renewable You

"The highest form of research is essentially play."
–Neville V. Scarfe

You want more people in your green organization, but you've been running around trying to do everything, draining yourself, and wondering why you can't find new people to join your group.

Energy matters and, in this case, your energy matters. Start by renewing yourself so you'll have the energy to renew others, bring more communities and diverse people into your group so that, together, you can make a difference for the Earth.

Play is a great path to renewable energy.

Puppy Power

Think of little kids, how they'll walk up to another little kid and say, "Wanna play?" Or think of a family of small puppies, how they roll around and wrestle each other. They embody the energy of play—and that is the energy of connection. It's the energy of connecting with ease. When you emulate this playful connectedness—whether you are an extrovert or introvert— you can connect more strongly with your community, or with someone new.

When a person watches another person happily playing with a puppy, mirror neurons in their brain start to fire together, meaning the observing person's brain starts to imitate and actually really feel what the person playing with the puppy is feeling. This can be wonderfully helpful to know and to be aware of. Mirror neurons fire when you do something, or when you watch someone else doing the same thing. So, when you smile and bring a relaxed, connected energy to meet someone, their brain will tend to imitate yours, and they will tend to feel a similar way.

Discover *Your* Way to Play

Play—what's the feeling of it in *your* body? Play can cover a lot of ground—from wandering around as a kid while playing in the mud, to playing a musical instrument, to participating in organized sports, with teams and specific rules.

A New Kind of Portrait

What we're interested in today is your personal play portrait. What feels playful to you? There may be many different things.

You can use this exercise to help you choose what will give *you* the most playful bang for your buck.

Imagine yourself going back, gently back, to a time when you felt especially safe and happy (or maybe this is a way you feel at times in your current life). You may have been a young child or a carefree teen. Perhaps you were walking along a favorite path, or hanging out near a beloved tree.

Bjorn enjoying mud-puddle play

What did you do for fun? What made you so happy that you forgot the time, and simply enjoyed what you were doing?

Once you've relaxed and have an image in your mind of an activity you totally loved, check in with yourself for more details about how you feel and what was happening. Allow yourself to sink back into those memories and enjoy them.

After a bit, gently allow yourself to return to here and now. Where were you? What did you discover?

You just rediscovered how you like to play.

This exercise can help you to renew yourself when you are stressed out; and get creative about what kinds of events and activities might attract a variety of people to your green organization.

Your Personal Playlist

Now it's time to write down your very own playlist—your go-to list of how you like to play. Take a snapshot of your remembered experience by writing it down or by finding a picture that represents it for you. Jot down some notes to help you remember it. Here are a few prompts to get you started:

- Were you in a favorite place? Outside? Inside?
- What kinds of things were you doing? Was there a certain activity you really loved?
- Did you play with others, or enjoy time to yourself?
- What toys or books did you play with?
- Were you running, jumping, dancing or exploring?

This is your own personal playlist. These are specific activities and ways you love to play and rejuvenate yourself, tap into your creativity, and connect. These are clues and hints about the things you love and the ways you love to play. They may have morphed somewhat as you grew up, but their essence remains.

Look for themes and specifics: "I loved being outside, running with my friends, and biking with my cousins." "I was a bookworm, always happily snuggled in a corner with the latest Nancy Drew mystery story." "I loved baking, especially baking cookies with my aunt. We sang songs and had such fun together."

Obviously, as a grown-up, you have added to your list of fun things. This exercise is an excellent way to start reminding yourself of your personal playlist.

If you're feeling shy about writing down "making paper airplanes" or some such activity that now feels too childish, remember that we're talking about play for its own sake. There does not have to be a "good reason" to do it. It does not have to be fancy or something you think you "should" like. It's what *you* enjoy. It's simply something you love to do.

Reinvigorate Your Team

I invite you to do this playlist exercise with your team to find out what people love to do. Draw a list of cross-connections (get out those colorful markers and a big piece of paper) and see where your team's playlists overlap. Get inspired from these overlapping playlists. This is like mind-mapping meets your playlists. Then get inventive and make your next team meeting creative and inspiring for your team.

You have four people on your team who love the outdoors? Oh, wait, it's your entire team who loves the outdoors! Great! That can turn into having a picnic at the park/in the woods/by the lake/near the bluffs + strategic planning + new social media ideas = successful team meeting.

Play Pockets

Keep your playlist handy in a pocket (this could be an electronic pocket). When you find a bit of time, five to ten minutes or so, say on the way to work, or on a break, insert some play into this pocket of time. Pick an activity from your playlist—something that makes your heart say, "Yes!" or makes your body relax—and do it. If you're not used to it, start slowly, and do it for five to fifteen minutes. You can enjoy the same activity again next time, or mix it up and try others. Play Frisbee, tell your friend a corny joke, or play with your dog.

Enjoy putting these play pockets into your week. Play, *simply play*, for the sake of enjoying yourself, at least three times this week. If you can insert a play pocket into every day this week, that's great. See what happens and how you feel. Make notes about what makes you happy and gets you feeling rejuvenated. Being rejuvenated allows you to have your best ideas, inspire your team, and make connections with diverse people more easily.

Practical Play

Let's take a look at how a playful attitude or approach could help your team with some of the challenges facing organizations, and especially non-profits.

A Creative Approach to Challenges

Let's say I was working with a client and she was struggling to bring in grants. Ling feels passionate about her organization, loves their mission, and wants to benefit more people, but is just not connecting the dots in finding grants. I asked her if she had ever chatted with a grant before, just to check in. She laughed, which is always a good start.

I invited Ling to imagine having a conversation with "Grant," a guy who represents grants. Let's say she meets him at a work gathering at the office. "So, you're at the office, chatting with co-workers, and someone introduces you to a guy named Grant. What is Grant wearing? What do you talk about? How are you feeling?"

Ling answers with, "He's got a suit on. We are making small talk. It's awkward. I feel stupid and like I don't know what to say. It's stupid office talk, and I feel like a fraud. I feel like running away."

Okay, that's some interesting information. With this "Grant," Ling is not able to talk or communicate, and feels awkward enough to actually feel like running away. She cannot communicate with this potential source of support for her organization.

I invite her to imagine another scenario where one of Ling's friends invites her over to her house or apartment, a place that's comfy and welcoming. There are a few friends coming over for a casual summer gathering. Ling's dog goes up to greet this new guy she hasn't met before, and she starts talking about the dog, and sharing goofy animal stories with him. When I ask Ling how she feels now, she says, "I feel great, comfortable. I'm laughing. I'm curious about this guy."

"Cool," I say. "His name is Grant."

She laughs, surprised, and then becomes thoughtful.

Shaking Loose New Possibilities
"What would it be like to feel comfortable and laugh and be curious about the grants you're writing?" I ask Ling.

She spends some time playing with that idea, which results in Ling applying for grants with an approach of curiosity, a feeling of being comfortable, and using playfulness to help her relax and focus. She gets into a place of curiosity (*What can I learn from this grant?*), comfort (*Do I have my favorite grant-writing mug with me?*), and playfulness (*How can I have fun writing this?* and *How can I communicate that our organization is creative and resilient?*)

She now has much more fun writing grants. Now she imagines her new, dog-loving friend "Grant," and writes the grants as a letter to him.

Clearing a Path Through Overwhelm

There are so many choices to be made when you're part of an environmental organization. You may wrestle with questions like, *How do I know which option is the best for our team? How can I maximize my effectiveness when there are a lot of ways to do this?* For example, which of many social media choices are the best to use when creating an invitation for a Latino/Latina group? We have so many qualified candidates; which person would be the best fit to add to our team? Which totally fun picture are you going to choose for your next Instagram photo to best reflect your team's and volunteers' accomplishments?

Making Great Choices

Another client, Alexis, was getting frazzled and overwhelmed by how many choices she needed to make on a daily basis. She works with bird refuges around the Pacific Northwest to ensure that they are protecting and releasing as many birds as they can, sharing their knowledge with the public, using best practices, and connecting with each other. This involves a lot of complex choices, and Alexis was having a hard time deciding between which choice was really good and which choice was great.

We decided to take a small step, and practice with a smaller decision. Alexis enjoys reading and was deciding which book to take along on a camping trip. She had been given a book by a good friend—an interesting, topical, highly relevant book. And yet… another book, about the triumph of the human spirit in Malawi, also caught her eye.

We used this situation as an example to help her tune in to a way to make great choices versus choices that are merely good.

A Game of Warmer/Colder

I invited Alexis to remember the warmer/colder game she's probably played as a kid. She giggles, and says, "Yeah, I remember that. I'd play that with a friend of mine. She would hide something while I wasn't watching. Then she'd give me clues about it to help me find it. If I was far away, she would say, 'Cold. Oh, you're so cold it's like ice cubes in here!' Then, as I moved around, trying to find it, she would say, "Cooler, cool, still cool," but as I got closer to the object, she would say, 'Warm, warmer, yeah, warmer yet,' and finally, as I got really close to it, she'd get all excited and say 'You're very warm, hot, so hot, lava hot!"

"Remember the feeling of excitement when you got your first 'warmer' clue?" I asked.

"Yeah," Alexis said, "I got all excited, and curious. And I felt *so* happy when I found the toy! I'd get all giggly!"

"That's the same feeling as a 'warmer' decision for you, or one that indicates a "Yes" for you, I said. It generally has your body feeling excited and happy, or relaxed and happy. In contrast, how did you feel with each successive 'colder' clue?"

"I felt kinda draggy, like my body got droopy," Alexis replied.

"That feeling of tension, or having less energy, is the feeling of 'colder' in your body, and indicates a "No" for you, I explained."

Alexis took a look at her two books, one at a time. Her body felt tight and tense for the first book, the "oh-so-topical and relevant" book, but her body relaxed when she held the second book, the "triumph of the human spirit" book. That's the book she gladly chose to take along on her camping trip.

The Warmer/Colder Decision-Making Game
You can use this game to see what's warmer or colder for you, by tuning in to your body, and allow it to help you make decisions.

Imagine a scenario for a few moments or minutes in rich detail—for example, having a meeting at a local school that has a community garden—and see if your body feels relaxed or if it feels tense at the idea. Then imagine a different scenario—for example, having the meeting at a local community center instead—and see what happens. When your body feels relaxed, that's "warmer", and means yes. When your body is tense, that's "colder" and means no. It'll take a bit of practice, but soon you'll notice yourself tuning in to your body's responses quickly, helping you make better decisions.

Renew Yourself So You Can Do Your Best Green Work

Your energy and the way your body feels matters, so take the time to build your energy up through play and feeling "warmer" and renew yourself. Your volunteers, your board, and your community deserve a green leader who is resilient and responsive. You deserve to feel renewed. Bend like the willow, stay strong like the oak, and keep your roots happy.

Sprouts: Your Green Take-Home Tips

Ask yourself, when faced with a challenge, Can I have a creative conversation with the challenge I'm facing? (Like Ling in the example with "Grant"). If this challenge or project were a person, what kind of chat would I want to have?

When you're weighing different options, ask yourself:

- *Does this decision make my body relax (warmer, meaning Yes) or make it tense up (colder, meaning No)? What other scenarios might work better?*
- *How can I renew myself daily?*
- *What types of activities would invigorate our organization?*

Next we'll explore how animals can inspire us through their natural playfulness.

Chapter 2
The Power of Animals and Play
Of Polar Bears, Crows, and Horses

"Maybe it's animal-ness that will make the world right again: the wisdom of elephants, the enthusiasm of canines, the grace of snakes, the mildness of anteaters."
–Carol Emshwiller

L et's take a look at our friends in the animal kingdom to see what inspiration and ideas we can learn from them about play and diversity.

Dogsled musher Brian Ladoon thought his sled dog, Hudson, would soon be dead when a polar bear approached the dog. Hudson was staked a bit away from the camp, along with the other sled dogs. After four months of hunger and waiting for the sea ice to freeze so it could go and catch its regular prey of seals, that bear was *hungry*. It was capable of a skull-crushing bite or a deadly paw swipe. The 1,200-pound polar bear could have killed the dog quickly.

Instead, he responded to Hudson's tail wag and bow. And they played! They played with such abandon that the polar bear had to stop and flop belly-up in the universal gesture of "I need a break." Brian, watching from his camp, was astounded.

The bear came back for another play date the next night. Soon the word spread, and people gathered at Ladoon's camp to watch this unusual pair play. At one point, the polar bear even enveloped Hudson in a gentle embrace. The huge bear came back every night for *seven* nights to play with his new friend the dog. Finally the sea ice was frozen and the bear was able to return to his usual hunting grounds for seal.

Brian kept the experience quiet for years, fearing that people would not believe him, and indeed, some of the first reactions were ones of disbelief and fear, but now you can see the pictures here: www.npr.org/sections/krulwich/2014/03/01/283993033/polar-bear-flip-flop-people-hated-then-loved-these-photos-what-changed.

This is a prime example of a totally unexpected connection, one that brought joy to three different species—bear, dog, and human. This true story shows you just how important play is for animals.

Bob Regan, a dedicated and well-respected field researcher, has watched grizzly bears at play for over a decade. He says that the most playful bears are the ones that do best in life. Their capacity to explore and discover their world through play is what allows them to have the most inventive responses to something new in an ever-changing environment. These bear stories are some of my favorite highlights from Stuart Brown's book on play (please see the resources section at the end of this book).

Leaping Across Huge Barriers

Maybe the realities of racism, classism, or ableism loom large for you. Maybe, when you think about reaching out to other communities, it feels like reaching out to a polar bear to invite her to tea. You fear that your invitation will be wrong, or seen to be wrong, or will be misinterpreted.

What if that big bear (or the group of people you'd like to invite) just really wants to play, to be able to participate with you? What if they simply want to be included and welcomed? Can you have the courage and the playful spontaneity of Hudson and go ahead and invite them (or respond to their invitation)?

We can learn so much from our animal friends about how to be more resilient, more playful, and even how to cross boundaries that seem impossible. Our capacity to play can stretch across borders and boundaries of race, class, religion, and language.

Animal Play and Its Advantages

Animals play in ways they *love* to play. For example, river otters love to twist and turn and poke their noses into things and explore, they swim like mad, and they slide down riverbanks and snow banks. They also run on land. They are flexible creatures in mind and body. They romp!

Animals you would not expect to play break all the rules and go for it. Huge, fully-grown bison, weighing from half a ton to a full ton each, will skid around on frozen lakes, bellowing. There is no "reason" for this that humans have identified. The bison are not going off to find food or mates. They seem to just want to play.

Corvids, crows and ravens, are famous for their playful inventiveness. A crow was hopping and pecking while trying to get at food. She accidentally fell over and slid on her back down a snowy slope. She then ignored the food and purposely slid downhill on her back on the snowy hill—over and over again, with apparent enjoyment.

Why do animals play? (Remember that we are animals, too.) It allows animals and ourselves to be more flexible in our responses to an ever-changing world. It actually confers an advantage to us. Not only is playing just plain fun, it is extremely useful in helping us find success in our day-to-day lives. So, let's take a closer look. What can you learn from the playfulness of animals?

Animal-Watching for Fun and Other Benefits

Is there a favorite animal in your life? Observe your animal friend carefully. Or go out into the natural world to watch the animals there.

What kind of a character is this animal? Does it seem curious, quiet, and observant? Or bold, brash, and friendly? How does it interact with new things in its environment? Or do you see a group of wild animals? If so, how are they playing?

Take some quick notes—gather the broad strokes of their personalities and characters. How can you use those characteristics to solve a current problem you're working on? For example, how could you be more observant? What could you be curious about, like the chipmunk you watched earlier was? How might that help you and your green organization?

Fudge the Horse and Mr. Peacock

Fudge, a horse who was my dear companion for more than 27 years, was a funny, smart, kind mare. One day, she was dozing out in the pasture, close to the barn, with one hind leg at rest. Suddenly, a big bird landed near her, and she startled, and snorted, and turned around to look. *What* was *that!?,* seemed to be what she was thinking. The bird was Mr. Peacock. Fudge snorted at him and walked toward him, as if to assure herself that it was, indeed, the peacock she had seen many times before and had no worries about.

Mr. Peacock flew up to the top of the milk house nearby, and Fudge settled back down. Once Fudge was again peacefully situated, ears at half-mast, mostly asleep. Yep, you guessed it, Mr. Peacock flew down and startled Fudge again! That time,

Fudge snorted and ran and bounced around, and Mr. Peacock flew away. But I swear he was laughing and saying, "Gotcha again!" It was *so* funny.

What could a humorous approach do for your next environmental work gathering? Or how about a surprise? What could you bring to a "sleepy situation" that would liven things up? Maybe your board meetings (read "bored" meetings) are about as exciting as old toast, and could use an infusion of surprising humor. How could you introduce an element of liveliness and laughter to your next project?

Trust in Surprising Places

Wheeeee! What fun it is to slide down... the head of a whale? In the ocean near Hawaii, a bottle-nosed dolphin was videotaped sliding down the side of a humpback whale's head. The action had no apparent evolutionary reason other than play for the sheer joy of it. It involved much trust by the dolphin, as well as cooperation from the enormous whale. Here's the link to this short, but amazing video that shares photographs of multiple instances of this playfulness: www. natureworldnews.com/articles/15712/20150718/dolphins-riding-whales-exactly-new.htm.

It will take trust and cooperation to involve other folks who are members of different ethnicities, religions, or worldviews, to invite them to join your green organization. When you think of your next green community project, remember that whale and dolphin and how they trusted each other and were able to create a wonderful, playful connection.

Animal Play

Next, you're invited to meet your playful animal mentor, through a fun guided visualization that you can access at *www.foolwillingbook.com/gifts*. This will be a fun way for you to get more ideas. Enjoy!

Move Your Body, Refresh Your Mind

Animals are usually moving when they are playing and of course, wild animals play outside. Movement is another way to let our creativity loose. What better way for someone who loves the planet, you wonderful greenie-granola-environmentalist you, than to move your body outdoors, on the earth you are working so hard to protect?

Move your body in a way that feels fun and invite others to join you. Have you ever noticed that you feel better after you've been moving playfully? Maybe you played a game of tag with your niece and nephew, or went hiking in a local park, or took your neighbor's dog for a walk and noticed that you felt refreshed afterwards? Just like for other animals, movement helps our brains stay creative. Movement outside helps us feel more connected.

I am not going to suggest that you go and slog in a boring gym where you stare at concrete walls in order to move. No. Ick. (Unless you like that, then go for it.) We humans are physical creatures and our play is often physical, but pay attention to what type of movement and what type of environment feels good.

Playful movement can immediately connect us to other cultures and communities. For example:

- Think of the gorgeous rhythms of salsa dancing and how the dancers dress in beautiful clothing.
- Imagine hands on drums, pounding out a catchy beat, celebrating a wedding in the African country of Ghana, the people nearby laughing, dancing, and celebrating.
- Imagine children from around the world and how robustly physical they can be—jumping, twirling, wriggling, and laughing.

Fun *and* Green

I invite you to move your body in a way that feels *fun*. Invite others to join you, and combine fun activities with green events. Here's a list of fun activities to try:

- First, host a community clean-up of a playground, then play flashlight tag in it when you're done.
- Do a hula-hoop dance competition to raise money to clean the lake in your neighborhood.
- Get to know your neighbors by hosting a photographic treasure hunt in which people look for (and take pictures of) things in the environment, like specific flowers and trees, any dog, someone on a bike, someone recycling, etc. They can enter a competition to win a prize.
- Have a unicycling parade to celebrate local endangered species.

- Sponsor a Ping-Pong© tournament to bring attention to a local injustice. Proceeds could go to benefit the cause.
- Walk with others in a park to enjoy the spring wildflowers or the fall leaves, and set up a treasure hunt for the young ones.
- Do a Weird Trash Contest—a recycling and garbage pick-up event that includes a contest to see who finds the weirdest piece of trash.
- Organize a Bike-n-Picnic event. Pick a favorite trail and invite participants to pack a lunch to take along. Or bike through a neighborhood so the kids can join in, and then eat the picnic lunch at their playground together.
- Do green street theater.
- Play animal charades at your next team meeting.

We move because it's good for us—good for our bodies, minds, and spirits. As a bonus, it's also good for our creativity. It's good to get away from the desk and back into the outdoors, our arena of celebration.

How can we harness this energy in a way that is playful, wonderful, and celebratory? Many cultures include the earth as a central part of their folkways. Explore ways you could partner with them to celebrate their green work and lifeways. Turn something mundane but important into something fun.

Here's an example of someone doing this. My client John had an important test to take for a certification. He knew the material, but, as a mechanical engineer, felt like he had the pressures of *Gotta be the best student* and *I must get this right!*

hanging over his head. Instead of preparing for and taking the test with the high-stress level that he normally would have, he decided to take a playful approach. He told me, "I just decided to play with it." That attitude made the test-taking much easier for him. Afterward, he said, laughing, "I even enjoyed it. It was fun to see how much I had learned." John passed his test with flying colors, high compliments, and a new way to do things in the world.

Be a Fool Who's Willing

When I was in my teens, I took a workshop to become a 4-H camp counselor. I was going to be the co-leader of the Music/Dance sections and I was really nervous. I'd gone to camp every year since I was eight years old, and I'd loved it, but it was going to be my first time as a counselor.

A blond, curly-haired, curvy woman talked with us, a group of awkward and eager teens. The first thing she said was, "You need to be fool-willing. You have to be willing to make a fool of yourself, and not mind, in front of others, so you can teach them. Then they will trust you and be willing to make mistakes in front of you, and they will learn from you." That moment, and those words of wisdom imprinted solidly onto me, and I have always remembered them.

I have allowed myself to learn and to teach new things in front of others, while repeating to myself, *Be fool-willing.* I have been thanked multiple times in lots of different situations for my willingness to be vulnerable, to stand up and try something new. I have laughed louder at jokes and funny situations because of being fool-willing. That statement, "Be fool-willing," has allowed me to be more myself.

Sometimes you might need to play the fool for yourself.

Jeanette had a big work move that she was dreading. She was moving her work to a new site and had many clients to contact and moving details to orchestrate. In the midst of all of the changes and frustrations, she remembered our coaching work together and made silly faces at herself in the mirror. She totally cracked herself up by exaggerating goofy faces. After being able to laugh at herself, Jeanette made the calls she needed to make to ensure that the transition to a new work site went smoothly for herself, her clients and colleagues. Her own playfulness had relaxed her to the point of doing needed tasks much more easily and with less angst.

Clownin' Around

When I was a pre-teen, I was at a weekend church camp with a bunch of early-teen kids and parents. Many of those kids were much more into flirting than I was and they were being kind of crazy, in the way that pre-teens can be, and I felt really lonely and left out. I felt uncomfortable joining in with kids I did not know very well, and who seemed rather out of control to me. It was a *loooong* and lonely weekend. Then, on Sunday, the very last day, I was given a choice to take a workshop about white face and to learn about clown make-up, its history, and even create and put on my own clown face. I could have gone with another kid I knew to a workshop I was not interested in, but I decided the clown workshop sounded fun, and I wanted to do at least *one* fun thing just for me in that whole, long, miserable weekend.

At the workshop, I learned that each clown takes the time to create his or her own unique face, and that people will tell clowns things that they will not entrust to anyone else. There's something sacred there, a trust that is dear to the clowning world. That flung open doors in my mind. I was intrigued by the rich history of symbols combined and used to create each unique clown face.

I created my own clown face, slowly applying each color of the thick make-up to my young face. At the end of the workshop, they said I could wash it off, but I didn't want to. I wanted to keep my new face, and even show it off.

A New Face

I kept my clown face on and the kids who I had come with, who had not paid much attention to me, suddenly did. By then I felt free and did not really care if they noticed or not. I was happy with me. It felt like a bonus that they were happy with me, too.

On the way home, I decided to share my new me with the world. I sat on the outside passenger seat of the big van we were traveling in, and looked out the window to surprise travelers passing by on the highway. I gave them my biggest grin and a big wave. Most people loved it, kids especially, and they waved and waved and exclaimed with laughter and enthusiasm. It felt like a gift, to allow this aspect of myself to be seen, to give the gift of laughter and surprise to total strangers.

I invite you to be fool-willing, to see what you can create from a willingness to be vulnerable, to really be seen in the world.

Discover what you can generate for your green organization from a place of playful vulnerability and responsiveness.

Humankind is made to play, and that crosses *all* kinds of boundaries.

Sprouts: Your Green Take-Home Tips

You'd love to perk things up in your events and meetings, to create engaged gatherings where people are really participating.

- What ideas came to mind when you imagined combining an earth-friendly event with outdoor play? Present some of those ideas to your team as a jump-off point, and brainstorm together about something your group can plan and do.

- Is there an animal that you love to watch play? Go outside and observe that animal, and/or go online and look for some videos of that animal at play. Take some notes about what characteristics their playfulness has. How could that type of playfulness inspire your next meeting? What different options do you have, or can you create, for including more of that quality of playfulness in your activities and your green organization?

Music crosses barriers of language, explores culture, and touches our hearts, all without words; how to use music to enliven your green organization is what's coming up.

Chapter 3
The Power of Music
Notes to Create Community

"I believe in kindness. Also in mischief. Also in singing, especially when singing is not necessarily prescribed."
–Mary Oliver

Music is a way humans can connect across multiple barriers. Music does not care what religion you are, or what your sexuality is, or that you forgot to return that library book again. Music reaches around and through those things. We connect with our hearts to the love and joy found in music.

In this chapter, we start with the first note, with your note, and we create joyful sound, then build that into a song and, lastly, create a choir—a choir that can make a difference for this sweet Earth, our home.

Uncover Your Musical Note

If the world was one great choir, what would your particular note sound like?

Your note is the compilation of your knowledge, experience, and—something even more important—your heart song. It is what makes you glad to be alive and what calls you forth.

That is a poetic way to describe it, but there's also this: The beating of your heart can be measured by something called a *magnetometer*. It is a scientific instrument the size of a coin that can measure the magnetic energy of your heartbeat from three feet away. So, yes, you really do have your very own personal heart song.

Your heart will beat more strongly when you are happy and fulfilled.

Drumroll, Please

Why are you here? This great question awaits your answer.

Imagine sitting down to tea with your heart. There are comfortable chairs or colorful cushions, and you're deep in conversation. Ask your heart, "Why am I here?" Repeat the question, diving deeper with each answer. And then you'll know. Here's an example:

- Q: Why have I come here, dear heart? A: *To help my family.*

- Q: And why is that? A: *Because I love them.*
- Q: Beneath that, why? A: *Because I want to feel needed.*
- Q: And, again, why? A: *Because I want to feel loved, and I want to give love.*

Diving Deep

You may start out with answers that are about money, or comfort, or making your mom happy, or your grandpa proud, and then, as you dig deeper, you will discover more and more profound reasons for why you are here.

"It's because I love my son and I want a better world for him." "It's because I love my grandpa, and I want to honor his legacy." "I love orangutans and want them to be well and happy." "I love the river waters near my home. I love the river. I want to help my beloved river."

This, my friend, your deep answer, is *your big why.* Tape it to your wall, your desk, or your forehead, if need be. It will call you forth, in times of difficulty or trouble, in a way nothing else will.

Whew! *Good work.* That was a *big* one! No worries if it's not exactly clear yet, or if you are still unsure. This is an exercise you can return to. The answer tends to deepen over time. For now, if you have a little inkling, or a nudge, that is good. You can expand on that. You can follow that nudge.

Here's the thing—as my professor Gretchen Schoff asked—"Is it a path with a heart?" It does not matter so much which path it is; what matters is that you can put your heart into this path, and that it feeds you as much as you feed it.

Music for Renewal

Now we'll dive in to the fun pool of music and simple sounds that unite us as humans. When you discover what sounds and music enliven you, you can use them when you're tired or drained, to help renew your heart and renew your purpose. You'll also see how you can use music and other sounds as a way to connect with people, even across boundaries of language, because music speaks with its own tongue.

Of Kazoos and Whoopee Cushions

"Make a joyful noise unto the Lord." Even in the Bible it says to make music, to make noise, simply noise, but to make it joyful. You can see this happening in many YouTube videos of tiny three-year-olds singing with enthusiasm and joy. They may be off-key and making up the words as they go, but they are singing with great joy!

I invite you to find something that makes a sound that brings you joy or a sense of playfulness. You can go for a sound that makes you feel silly or goofy. Maybe you have always loved kazoos, and think they are so funny. Or the noisemakers that come out for the New Year. Or slide whistles at birthday celebrations. Maybe there's a sound in nature that cracks you up—the sound of a dog yawning (*yaaaaw-Rup!*), or a bird call that's really funny. You can find a wide, wide variety of silly sounds on the Internet, or ones already on your smartphone. Check in with a five-year-old who you know, and request some silly sounds, record them, and you'll be set. This does not need to be fancy, or even particularly polite –you might get a big kick out of whoopee cushions!

Sound Signals

I'd like you to use your chosen, playful noisemaker to signal to yourself that it's time to tune in, to relax, and to get playful and creative. Here's a few ways you can use it:

- Remind yourself to take a break.
- Ask yourself, How can I bring more playfulness into this moment? What can I do to enjoy this moment?
- Signal to yourself that it's time to let loose, to let the ideas fly forth, to write (not edit), to laugh, to make silly comments.

This is a sound to signal a time for inventiveness and creativity. It can also be used during team meetings to crack people up (a duck quack, for example), when you want to energize people, or to help them to relax.

The point is not to startle people, but to help them to laugh, relax, and access their creative selves. Laughter connects people and helps them know they're in a safe environment. Relaxed and happy people generate better ideas and do better, more creative work.

Personal Playlist—The Musical Kind

Another way to use sound to connect playfully is to have your own personal play song—one that lifts you up when you're feeling down, or simply makes you laugh, or feel goofy and delighted. I love the Muppets. I love their silliness and irreverence. When I sing or hum the Muppet theme song (and

I do!), it helps shed light on my difficulties, helps me laugh at myself, and moves me forward.

Look for a song that you love. It could be a longtime favorite, something brand new that speaks to you, or something you invented yourself. Maybe it is one you wrote with your garage band in high school. Hum it to yourself, belt it out in the shower, and put it in your phone so you can listen to it easily. Turn to this song when you need a lift.

For the music lovers among us, this may turn into an album! For the purpose of this exercise, pick one song to start with that you know lifts your spirits. Listening to it will help to tune your brain, quite literally, to the tune of happiness, playfulness, and creativity.

This is a fun group exploration, as well, for the people in your office or green team or mom's group. Invite each person to bring their play song, or a couple of choices from their playful list, to share. Compile an album for the group. It will be fun and revealing.

Lisa, one of my clients, heard about the popularity of the "Chicken Dance" here in Wisconsin. (We even dance the "Chicken Dance" at weddings! I know; we *are* easily amused.) She wanted to try it, and wanted to see if her husband would be willing to dance with her. Lisa found videos on YouTube and did the dance moves, giggling her way through it. After she practiced a few times, she invited her husband to join her.

His first reaction was "What on earth are you doing?" When she delightedly replied, "The Chicken Dance!," he laughed, and was persuaded to join her. They had a blast.

Lisa was willing to try something brand new to her and through this goofy "Chicken Dance," connected with her husband in a way that was playful and rejuvenating for both of them.

The Transcendence of Music

Music helps us move beyond our daily reality, helps us have a wider view, and can give us relief, make us laugh, and encourage us.

The Voice and Beyond

Music is so popular, as seen by the wide variety and number of TV shows that highlight people singing on the national stage for the first time and people just learning how to dance. Think of shows like *American Idol*, *The Voice*, and of *Dancing with the Stars*. These shows appeal to thousands of people—across racial, economic, and social lines. Producers of these shows are smart and have made many of them interactive. You can vote for your favorite performer from your living room.

Remember the delight of discovering new voices, and dancing? We watched dance practices, caught our breaths at the stumbles and the falls, and cheered people on as we watched them pick themselves back up and continue to practice. Remember the surprise of finding a brand new talent, their stories of the courage and the work they'd put in, and their joy of practicing? Imagine bringing more of that kind of courage, of trying something new (in front of people!) and that kind of joy and pride to the green movement. (It's already here. Let's expand it!)

Right now, perhaps you are tired of the same ideas, the same events, the same faces (even though you appreciate them), simply more of *the same, same, same* in your green organization. Imagine inviting new faces up to the front of your conference room, with their new and practical ideas for how to solve that seemingly intractable environmental problem facing your community.

You can bring more attention and creativity to your events and conferences through music and entertainment. We simply need to integrate experiences that humanity craves—like music, joy, and connection—into what we're already doing.

Making New Music

"*It's New Music Thursday!*" the local DJ announces in a thrilled voice when s/he showcases new music from an artist, or when there's a new band in town. *New Music!* The radio station has special times when they highlight the new music or talk with the band members. It's a big deal, and there's much musical creativity to celebrate. There's a huge variety of new music and songs being invented all the time—blues, industrial, Hmong pipes, reggae, and so much more.

When you go out into communities, you can invite them to share their music with you. This is such a lovely personal expression. Many folks will love it that you asked, and be delighted to share with you. Imagine asking an elderly Hmong grandmother what songs she loves. She, not understanding your words, turns and asks her granddaughter to translate.

Understanding then, and smiling at you, she starts to hum a tune you've never heard before, but it is lovely, and catchy, and you hum along, and smile with her.

Imagine being out in another community and going to a celebration with food and music, and learning their dances. They will discover that you are willing to try, that you are open to their culture, that you are moving your body to "their" music. This speaks volumes. This, my friends, is the start of community.

It is those moments of connection across multiple barriers that we connect as humans. We can start a conversation from there. Joining in with music or dance lets people know that you are friendly and willing. It helps people be more open to overtures of connection and the invitation to participate in your green work.

Musical Ponderings

So, how can you incorporate music into your current events? Can you create a new event, with music at its heart, which also has an environmental theme?

Invite another community to an event you co-create and highlight their music and their dances.

Here are a few ideas for event titles:

- Blues to Help Mother Earth Stay Blue
- Groovin' for Mother Earth
- Hip-Hop for the Planet
- Jammin' for the Planet

World Choir

Imagine a choir singing together when *something* transcends all of their voices. Their voices, united, become the song of the Universe, so beautiful that when you hear it—and then, later, remember it—tears prick your eyes? Yes, *that* kind of song, *that* kind of beauty. Allow that sense of joined voices, that beauty, to lift you, to remind you that you are a part of a mighty choir, a choir singing "*For the Beauty of the Earth,*" to renew and restore this precious place that is our home.

On the other hand are the song-eaters, the ones who prefer not to hear the music of connection, the destroyers and usurpers of our planet, our home. There are frightening characters called the "Echthroi" in Madeline L'Engle's wonderful book *A Swiftly Tilting Planet.* They are the ones singing and acting deliberately out of harmony, determined to destroy.

We all know the results of this—the lead in the drinking water of Flint, Michigan, that was shamefully and deliberately covered up, causing thousands of children to be exposed to the toxin lead, the toxic waste dumps located near marginalized communities; the slow and disorganized response to Hurricane Katrina. Then there are the climate-change deniers, the poachers of the wild and the beautiful, and the racist bigots who spread toxic sludge with their words. These kinds of folks try to destroy that the song of healing.

There is much room for many, many different kinds of music, and many different kinds of song, but we need people singing the song of healing, of kindness, of restoration for the Earth. Your "song" may look like helping to organize a lake clean-up, or speaking up for the local park or a spot where your neighbors enjoy bird watching, or helping a local school in a marginalized community to create its own school garden.

Diversity in community creates a symphony. When we reach out to communities, and when we bring in diverse communities—communities of color; lesbian, gay, bisexual, transgender and queer (LGBTQ) people; the Latino community, the Black community, Native communities; people of different ages, faiths, and abilities—*all* different kinds of people—to work together to solve environmental problems, we create a healing symphony, one that is beautiful and magnificent, and deeply affecting and deeply effective.

Diversity is our symphony of beauty, strength, and healing.

Sprouts: Your Green Take-Home Tips

Right now, your meetings may be very solemn affairs, very business-like and formal, and it may feel like the organization is stagnating. See what happens when you thoughtfully add music to your gatherings. For example, you can:

- Play music. This could be an introduction to the meeting, a way to send out a call for participants (when this riff is played it's time to gather, or time for a break, etc.), or even a part of the meeting itself.

- Encourage connections to form by playing music at your meetings, gatherings, board meetings, events, and conferences, as people mingle. Play thoughtfully chosen background music and watch how a roomful of people relax and connect, simply through the addition of music.

Next we'll look at words, and how we can use them wisely, to shift old biases, create connections, and strengthen bonds.

Chapter 4

The Power of Playing with Words
Wordplay and Thought Work

"Wordplay hides a key to reality that the dictionary tries in vain to lock inside every free word."

–Julio Cortázar

Words have power, such amazing power. They can unite or divide us. They can entertain us or scare us. They can confuse us. They can illuminate us. Words can be held up as a torch, as a signal of a way forward, or a marker of help, like the Statute of Liberty, who proudly bears the light for "huddled masses yearning to breathe free."

Let's learn to use words to free ourselves, to embolden us, to entertain and enlighten us.

Shakespeare, Facebook, and Maya Angelou

Wordplay is the "playful or clever use of words," and the first known use of the word was in 1844, according to Mirriam-Webster.com/dictionary.

Wordplay is everywhere—on Facebook, Twitter, and in so many other places on the Internet. There are Scrabble tournaments, daily crossword puzzles in major newspapers, and books and magazines offering many different kinds of wordplay. We are steeped in words, and we play with them.

We admire the people I call the Masters of Wordplay, like William Shakespeare, the poet Rumi, and the poet and word artist Maya Angelou. Words can become high art in a calligrapher's skilled hands. Words can be used as exhibition, as defiance, and as celebration; in poetry slams and witty one-act shows.

Words can also be a major tool for bringing together a wider community to participate in green activities that sustain our communities, and protect our land, air and water.

The Power of Ridiculous

Let's talk about a word that can stop us, unless we recognize its power. Then it can work for us. That word is *ridiculous*.

Ridiculous is defined by the Mirriam-Webster.com/ dictionary as "arousing or deserving ridicule, extremely silly or unreasonable." Many of us attempt to avoid feeling ridiculous, or being ridiculed, at all costs.

But if you are willing to risk ridicule, to step outside of your comfort zones, and outside of someone else's comfort zone, there is fire there, and juice, and the freedom to invent your life and your green organization the way you want to. This is how things can change; brave and kind people like yourself step outside their comfort zones, are willing to be "unreasonable" and "silly" and "ridiculous," to make the changes that need to be made in the world.

The Courage to Be

Who would you *be* if you were not afraid of being ridiculous? Invite yourself into your most courageous skin, and answer that question.

"What if I'm afraid someone will laugh *at* me? Won't that hurt? How do I not *care* if someone laughs at me?" Good questions. "If I am my true self, I will be ridiculed." We tell ourselves these terrible stories. And most of these are never true. This quote is often misattributed to Mark Twain, but was actually said by an anonymous octogenarian. It's succinct wisdom either way.

Maybe your answer will look something like this: *If I was not afraid of being ridiculous, I would be kind and fierce and brave and loving. I would speak up for myself and others who need me. I would throw the greatness of my heart into restoring this beautiful planet, in honor of all who live here.*

What is *your* answer to the question *"What if I was willing to be 'ridiculous' on behalf of something larger than myself?"*

Stories of Gold, Stories of Shadow

Many stories are gold. Stories can also be shadow. Stories help us define who we are as a culture, as a people, and as individuals. We write the stories of our lives in the choices we make.

This is powerful stuff, so I advise you to choose your story carefully. I also advise you to not take it *very* seriously, and to remember that you can change your story at any time. Have fun with your stories—the stories about your life, your organization, your community, the Earth—play with them, change the characters around, and write a brand-new, powerful, hilarious and meaningful script.

Let's take a look at the power of words through an exercise about story-writing:

1. **Write an overview of your organization's story.** Get yourself a pen and paper, or start a brand new document on your computer. Write the story of your environmental organization, or your personal green work, from the point of view of all that has not worked very well—all of the difficulties, the hard stuff, and the stuff that had you pulling your hair out. Give yourself five or ten minutes to write an overview.

2. **Rewrite the story, but with heroes and heroines.** Now write that same story, but from the view of yourself and your colleagues as the heroes, as the ones who overcame challenges and found inner resources—the "fools" who outwitted the kings and won the day.

3. **Compare the stories.** Which story *feels* more empowering? Notice that it's the same story, with the same facts, but the way you told it mattered immensely.

Think of someone you know who is delightfully witty and funny, and how the facts of the story they tell may be really simple, but the *way* they tell it makes it hilarious. Then think of someone who can tell about a light and inherently hilarious incident in a way that *feels* like drudgery to listen to. It could even be the same story.

The exercise above is a fascinating one to do as a group. Talk about what you discovered by doing it. See if there are ways, as individuals and as a group, that you can tell an empowering story about yourselves and your organization.

Pretend to Be the Director of a Movie

Sometimes, in order to create a new way to do things, you might need to set aside your own doubts that it's even possible, and willingly suspend disbelief. You can identify disbelieving thoughts by identifying the *yeah, buts* in your mind. Another way is to notice what things other people tell you are not possible, or just take a gander at the unhelpful or nasty thoughts flouncing through your mind.

For example, you might have floated out a new idea for how to increase diversity to a friend or colleague, only to be greeted with "Yeah, but that hasn't been done before" or "Yeah, but that's not the way we do things" or "Yeah, but 'they' won't like it." Similar thoughts may have landed in your head, or already been there and had you doubting.

Ask yourself to suspend disbelief. It's like you do when you're reading a great novel or are watching a gripping movie—you know it's not real, but you allow yourself to believe it, and you get deeply involved with the people in the story. You get absorbed in the storyline. When looking at the stories you develop with your organization and your own history, allow yourself the freedom to say, *Well, I know it's not real... yet*. We need to create a friendly, welcoming place for those new dreams to grow *before* they become reality.

Pretend that you are the director of this movie you're living, that you are organizing the characters and the plot line, and you're lining things up beautifully so they will work out.

When you engage in the willing suspension of disbelief, you engage the power to make something new. You give yourself a new stage upon which to tell a more engaging story.

Internal Biases

Internal biases are the hidden, or perhaps not-so-hidden, beliefs of intolerance and discrimination that can stop us and any kind of useful conversation in its tracks. Biases about which person "should be" a leader, or what an "ideal volunteer" looks like, what leadership looks like, can make us stumble.

Racism, sexism, ableism, ageism—you name it—all of those nasty "-isms" live in our brains as thoughts. We may have learned them from our school, religious institution, teachers, parents, schools, or elsewhere. The fact that these "-isms" are in your brain is both the good news and the bad news. We know they live in other people's brains, too, but *your* brain is the one you have to work with, so let's find out ways to uncover them and start to wiggle them loose.

Because, here's the thing, we *can* change our thoughts. Thoughts are literally electrochemical impulses that go through the neurons (cells that look like little trees) of our brains. It takes time, and it takes work to change thoughts—ours and other people, through our examples—but it *is* possible. And it's not that hard. It takes courage and willingness. And it starts with you. As the great song from Bob Marley says, "Free yourself from mental slavery."

Loosening Thought Tangles

"Thought work" is the capacity to, first, recognize your own thoughts, and then be able to use tools and exercises to work with and change your thoughts so you can create more empowering thoughts and patterns of thoughts. Thought work allows you more freedom, more capacity to recognize when you're stuck in a thought that's stopping you, and the ability to choose a different thought and pattern.

When you give yourself the gift of doing thought work, you can wiggle loose a bunch of disempowering, biased thoughts. If you do enough thought work, those thoughts will even dissolve, and you can choose new thoughts that free and empower you.

Byron Katie is a thought-work leader, and created her process called The Work. Martha Beck, my mentor/teacher, is also a wonderful thought-work leader. I highly recommend reading their work. (See the resources section in the back of this book for more information.). What I'll share with you about thought work is based on a combination of their work, with some sprinkles of my own.

You'll notice that, after doing thought work, your mind will feel clearer and more free. You'll feel better, and be able to see and create new possibilities, both for yourself and your green organization. Here's how it works.

First, identify the most horrid thought you can think of—the one that makes your gut clench, or makes you feel ashamed to admit you're thinking it. Yeah, *that* one. Then write it down. (It's okay. It will not actually burn your eyes to see it in writing.) Write it down and let it out, so that you can change it. You do not have to share it with anyone.

Now imagine that you're in a place where only the truth can be spoken. Everyone here is breathing fresh air made of truth. Think of a bright, clear day, or pure water, to help you.

Approach your thought with the most basic form of thought work, by asking yourself:

1. *Is this thought helping me or is this thought hurting me?*
2. *What new and empowering thought could I think instead?*

If you can rid yourself of nasty, annoying, prejudiced thoughts with that simple formula, that is awesome! You can quickly step back, notice what you're thinking, see if your thought is hurting you, and, if so, choose a new one. It may surprise you to discover how freeing this feels, how there is more freedom for you to choose your thoughts.

Going Deeper

You may still be feeling stuck, or not noticing even a little freedom or wiggle room around the icky thought. In this case, you may realize that you have old biases or prejudices that need a deeper form of detangling. (We all do; it's okay.) Now, to go deeper, ask yourself:

1. *Is this thought helping me or is this thought hurting me?*
2. *How is this thought hurting me? Get specific with some follow-up questions: How do I treat myself when I think that thought? How do I treat others? Do I act differently when I think that thought? (Do I get quiet, depressed, cynical, etc.?) How does my body feel when I think that thought? Is it warmer (relaxed) or colder (tense)?*
3. *Do I want to keep that thought? (It is a choice.)*
4. *What is a new thought, one that would move me forward? (Or, said some other ways, What new thought would give me more breathing room, more freedom in my world? What new thought helps my body relax?)*
5. *Choose a new, empowering thought, and discover ways to demonstrate how it is true/could be true in my life.*
6. *How is that new thought true in my life?*

Detangling Work in Action

I'll give you an example of thought work in action so you can see how it works. Before I do, I invite you to write down a thought that hurts you (write it in a simple sentence), so you can work through the exercise right along with Sydney. One caveat—do not write down feelings, as in *I feel sad/mad when x happens*, or *I feel upset about x situation*. Your feelings are valid; it's your thoughts we want to look at. Your thoughts may be causing those difficult feelings, so dig around underneath the feeling to identify the thought that is causing it. It may feel uncomfortable. Stick with it. It is totally worth doing.

My client Sydney's thought was *I'm not really qualified for good jobs*. Here's how asking herself questions and doing deeper thought work helped her get to a better place:

1. *Is this thought helping me or is this thought hurting me?* When Sydney asked herself if that thought *I'm not really qualified for good jobs* helped or hurt her, it clearly was a thought that hurt her.

2. *How is this thought hurting me?* Sydney's answer was: "I avoid my colleagues, and I tell myself that I'm not competent. My body gets tight and tense. I withdraw from my partner. I avoid applying for other jobs or opportunities."

3. *Do I want to keep that thought?* No! She did not want to keep that thought.

4. *What is a replacement thought, one that would move me forward?* She said, "I'm good at working with lots of different kinds of people. Good jobs require good

people skills, and I have good people skills. I can earn more qualifications if I choose to."

5. *Do I believe this new thought?* Yes, she did. It felt a little new, and wobbly, but she could feel that it was already helping her and, with more practice, it would feel solid.

6. *Choose a replacement thought, and then demonstrate how it is true in my life.* Sydney chose the replacement thought, *I'm good at working with lots of different kinds of people.*

7. *How is that replacement thought true in my life?* The ways her replacement thought "*I'm good at working with lots of different kinds of people,*" was true for her included: *I often get compliments on working with clients. I enjoy helping people solve a difficulty. I feel confident and buoyed up when I've helped someone with a problem. I love the challenge of helping so-called "difficult" clients with complex issues.*

New Possibilities

What is newly available to you with your new and more empowering thought? How do you now treat yourself? Others? How does your body feel? Check in. Are you breathing more deeply? Gently ask yourself, *What else is now possible for me?*

It may feel like you're out in new territory. You may stumble, and feel awkward. Or perhaps the new possibilities are spilling from your lips like a beautiful fountain. This is good news! You are literally building pathways for new thoughts in your brain. When you think a thought repeatedly, your nerve cells connect together and make those connections even stronger. It's like adding tiny wires together until they form a big cord. This is why it feels like you're "off-roading it" with the new thought; it's because you are literally building a brand-new nerve pathway with your new thoughts. The cool thing is that it only takes three times to think a new thought to have it form a pathway in your brain.

New thoughts allow for new actions, new strategies, and new ways of being. We can solve old problems with new eyes—eyes that have been opened by fresh thinking.

Thought-Work Strategies

Playing with thoughts can take different forms. Here are a couple more strategies to experiment with; try out different ones to see what works best for you.

Freed by the Opposite

You can ask yourself, *What's the direct opposite of this hurtful thought? How can I prove to myself that this new, opposite thought, is just as true, or even more true, than the original thought?*

It's good to give your brain other material to work with; it allows it to prove something else is possible, and gives you more breathing room. For example, let's take this particularly slimy thought: *Our organization is so stuck we will never invite more diverse people to join us, and we will have to close our doors.*

A good way to tackle this is to distill that thought down to its essence; for example, *We will never invite more diverse people to join us and, eventually, we will have to close our doors.* And, finally, to this: *We will never invite more diverse people to join us.*

What is the exact opposite of that thought? *We will always invite more diverse people to join us.* How does that new thought feel? Does it feel freeing? Do you feel more open, more creative, when you think that? Is it something you can follow up on?

Similar opposites can also be helpful. In this case, that might look like this: *We always invite diverse people to join us.*

Thought-Warming

Another strategy is to try warming the thought up a bit, if the full-bore opposite thought is too alarming to your system. Try to warm a new thought gently, like some delicious, home-made soup.

1. Original thought: *We will never invite more diverse people to join us.*
2. Warmed-up thoughts: *Sometimes we invite more diverse people to join us. Because we already invited more diverse people to join us, we have more diversity in our participants than we did last year. We can create more opportunities for diverse people to join us.*
3. Opposite thought: Now you have eased your way to the opposite thought of *We always invite diverse people to join us.*

Now we take the next step to allow your brain the chance to see how the new thought could be true, or just as true, as the original thought.

Here is the new thought: "*We always invite diverse people to join us.*" Then we ask, how is that new thought just as true or perhaps even more true, as the original thought?

Here are some possible examples of how "*We always invite diverse people to join us*" is just as true: *Our Facebook group is active and has a diverse range of people. Our spring event attracts more and more people from the Hmong and Latino communities each year. We have a translation service for three different languages, and have been using it regularly for three years.*

Your thoughts will vary. Keep digging and finding the ones that feel good. Look for thoughts that feel even better. *I am living in a sea of diversity, and it's easy to reach out.* How could that be just as true or more true? Here are some examples: *I just open my computer or smartphone, and there is diversity via the Internet. I hung out at my local library for a couple of hours, working on a project, and noticed people from four different ethnic groups. While shopping at the grocery store, I noticed another family laughing together and talking in another language.*

Keep uncovering and discovering these connections to new, better-feeling thoughts. At first, thoughts like these about diversity, for example, may seem small, but exploring them is how we begin to notice that a wide variety of people and experiences are really close to us, much closer than we think. When we notice this, we notice that inviting community and play into our green organization is not so crazy or hard after all.

The people to invite are right here, already in your life.

Lights, Camera, Action!

Another way to play this is to imagine your life, or the life of your group as a movie. Imagine the lead characters, the plotlines, the interconnections. What kind of movie do you want to write for the life of your organization? Drama, comedy, mystery, romance (this could be a romance with the Earth or a platonic love). How do you want to act? Bold, playful, and responsive? Or creative, quiet, and progressive? What do you want your character to do? Sketch the basic plotline. Write specific dialog. Have the characters problem-solve and go off adventuring. This exploration can show you exactly what you *really* want to do to increase the diversity of your environmental group.

To start exploring, finish these prompts:

- I could invite _____ to contribute like this: _____.
- She has great ideas; I could ask her about _____.
- Maybe I could chat with _____.
- There are many champions in this story, and their names are _____.
- In a surprising plot twist, the main character succeeds by _____.

A Golden Story for Your Organization

You can write a new and more empowering story for your environmental organization. You can imagine one that you'd like to create, and then to make real.

Take a bit of time, say 30 minutes, for this forward-thinking exercise. You'll want to let your inner seven-year-old out, the one who can imagine really amazing things. Set aside the inner critic/judge/cynic/inner snark for now. Let your imagination fly.

Pick a date, say three to five years out, and write what you'd love to see happening in your organization by then. Write down the date and then allow yourself to imagine…

- Look around. What's new or different or has shifted form?
- How are your meetings different? What kinds of events do you host or co-host?
- What communities of people are now involved, and have invested, in your green organization?
- How is your organization more diverse?
- What have you accomplished? What markers of success have you met? What are you celebrating?
- Have you recently launched something new? How's it going?

Here are more prompts to help you get started. (Use the ones that inspire you the most):

- You're walking with a good friend into a gathering of people. You are welcomed by _____.
- After a thoughtful, productive meeting, everyone gathers for refreshments and listens to music by a local band, playing at _____.

- The neighborhood folks know you by name, and love the project you co-hosted to clean up the kids' school and plant a school garden. Next steps together include _____.

- A series of events is planned. What's unusual about them is _____.

- Five groups of people are meeting for the fourth month in a row, and people have gelled. They work together as teams, and are thrilled about their progress together.

  They came together around the issue of _____.
  One of their first accomplishments was _____.
  Their action plan includes _____.
  Their next activity/event will be _____.

Notice what has changed in these five years. Where is the *biggest* change? Maybe it's "Oh, how we do business is fundamentally different" or "We're doing seed exchanges and plant trading as a yearly and much-anticipated event. Part of the event includes a Plants & Poetry word-slam, with an open mic for musicians." And, "Meetings are now held and celebrated at the Neighborhood House. There is much laughter and interesting discussion and things get accomplished right at the meeting. People go home and talk about the meetings with their neighbors, and attendance is steadily growing."

As you reflect back on this exercise, consider these questions:

- What was your biggest surprise?
- Was there anything you came up with that felt particularly appealing, like something that seemed like a wonderful idea, one you'd like to try soon?
- What were some of the themes?
- Did anything repeat itself from one scenario to the next?

Through this exercise of your imagination, you can create a brand-new story, and a whole new way for you and your green organization to do your good work in the world.

Sprouts: Your Green Take-Home Tips

Changing minds or writing new stories may, on the surface, seem like it is work that's not "green work," but if you're inviting new communities to the table, it is. All kinds of things are "green work." When the U.S. Clean Water Act was passed, that represented a "new" thought—that our waters, rivers, lakes, and groundwater, needed protection—and we rallied around it, celebrated our new thought, and protected these precious waters.

1. Doing your thought work helps you notice your own internal biases, which includes the language you use in your thoughts. This clears the way for new possibilities

for your community and organization, and for new ways that words can join in the fun as you organize for change.

2. Writing a "future-casted" story creates a powerful message to your brain that that eventuality, or something better, is actually possible.

Coming up we'll discover how the power of play can help you build and enjoy being part of your green agency.

Chapter 5
The Power of Play

"Play keeps us vital and alive. It gives us an enthusiasm for life that is irreplaceable."
–Lucia Capacchione

The gold-medal winner was a big, tall, Olympic champion, a member of a winning soccer team. He was a star player. But until he visited a little village a world away, he didn't understand how important a simple ball could be. There had been war, and armed conflict was continuing, and he was startled as he watched the kids play, because, in his words, "They all just played gun games, nothing else." They did not have any other toys, so they made sticks into guns.

The gold-medalist realized that the next generation coming up had nothing to play with better than sticks, and no examples of how to play, and that they needed something to give their young imaginations a different way to live.

He went back home, raised money, and bought soccer equipment—balls, nets, and other gear. As he traveled back to the village, he was nervous. *Would the kids like it? Would the village welcome it?*

Imagine the little kids—some bold, others hanging back—gathering around the tall stranger. And then he brings out a couple of soccer balls… and kicks one. *Ah ha! So that's what you do with it!* The kids loved it! He played with them to show them how to play the game, shared the rules of the game, and gave them some coaching.

He gave those kids not only a game, but a new way to be in the world, a way to interact in community without violence. He gave them a new possibility for their futures.

I read that story when I was in high school, more than twenty years ago, and the power of it has stuck with me. I would so love to know the reference for this true story and the name of the gold-medalist so I could mention them. (*If you know the names and details of this story, please reach out to me. My contact information is in the back of the book.*)

I include this story here because peace and war have everything to do with the environmental movement. War destroys resources that are vital to a community's basic survival—clean water, arable land for farming, schools, hospitals, and safe roads to travel. By giving those children a chance to imagine themselves as team players, and to engage in play that did not involve re-creations of war, the gold-medalist gave them a new way to envision and live in their world.

The power of play crossed lots of barriers in this village—language, religion, and culture.

There are many environmental ills that cross human-made boundaries of countries or cities. Wastewater flows into downstream communities; air pollution affects people thousands of miles away; habitats for bees that are destroyed in one area decrease essential food crops grown in another area because there are no longer bees to pollinate the crops. When we are working on multi-faceted environmental problems, we can use the power of play to make new connections and cross old barriers to help solve complex problems and challenges.

A New Lens

Many of the wrongs visited on our planet are also unfairly and disproportionately visited on people who do not have much political power. Often, that has been communities of color, indigenous peoples, and people without economic power. These barriers loom large.

The eco-justice movement is validating interconnections—between poverty and an impoverished environment, for example; between the empowerment of women and the empowerment of an entire community. Interconnections truly matter. We can use these connections for good, and to create diverse and accessible green organizations.

When we diversify our green organizations, we bring much wisdom to the table. We bring new voices that need to be heard. We bring different ideas, new intersections, and a sense of inventiveness that does not happen without this kind of cross-pollination. We can bring newfound hope, and new voices and energy to help solve serious and long-standing problems.

Let's look at our many environmental challenges through a new lens, the lens of playfulness and creativity. Like the kind Olympic soccer player, we can use play to leap over barriers and create community, new opportunities, and new hope. We can access our creative minds with ease, together, to come up with solutions that work for the community as a whole.

Play is the Thing

Let's back up a moment. What, exactly, *is* play? We'll start with a few simple definitions.

Play—"an absorbing, apparently purposeless activity that provides enjoyment and a suspension of self-consciousness and sense of time. It is also self-motivating and makes you want to do it again." This is from Stuart Brown, MD, who, along with Christopher Vaughan, wrote *Play: How It Shapes the Brain, Opens the Imagination, and Invigorates the Soul*.

As noted in Brown's book, some of the properties of play are that it is:

1. Apparently purposeless—You play simply because you love it; it is done for its own sake.
2. Voluntary—We come to it gladly and freely; no one "makes us" do it.

3. Inherent attraction—There is a sense of magnetism to it.
4. Freedom from time—We lose a sense of time.
5. Diminished consciousness of self—We are less aware, or not aware of ourselves as individuals.
6. Improvisational potential—During play, there's a feeling of openness, and there are ways to improvise and change the play.
7. Continuation of desire—We want to do it again.

Imagine having some of these appealing properties involved in your next work group at your green organization. Imagine people wanting to meet again, people improvising new solutions to a resource problem. Imagine a variety of people working together unselfconsciously.

Diane Ackerman includes this beautiful definition of play in her book *Deep Play*: "Deep play is: 1. A state of unselfconscious engagement with our surroundings; 2. An exalted zone of transcendence over time; and 3. A state of optimal creative capacity."

So now we have a few definitions of play. If you think about it, they include everything from quiet drawing of wildflowers in nature to fast running games played at full-tilt with friends.

Notice the qualities of "optimal creative capacity" and "diminished consciousness of self" and "improvisational potential." That sounds like an excellent recipe for a community of diverse people to work together to solve an environmental problem in their midst. Luckily for us, as humans we are wired for creativeness.

Neoteny and Your Flexible Brain

Watching kittens or puppies play is such fun. It's similar to watching a young child play; there's a lot of rolling around, and exploring the world, and finding out how things work.

Those young cats and pups will grow out of that playful stage, their brains becoming more fixed and less adaptable. For we humans, this is not so. Our brains have the function of *neoteny*, which means they remain flexible, capable of making new connections, all of our lives. Our human brain has great capacity to change.

This is why an 85-year-old baker can create a new cookie recipe that wows everyone, or a 70-year-old architect can design an entirely new building. You have this capacity as well.

That's a bonus for all of us, because it means our brains are malleable, are changeable, and play is a key part of this.

When we play, we invent new responses to situations—in the moment. We catch a ball in a new way, connect words on the Scrabble board differently, and this helps us come up with creative solutions to other parts of our lives—like how to hang up the coats in the entry way, or a new way to solve conflicts with your team, or how to clean up the illegal dump in our neighborhood.

This play, this ability to use our malleable brains to be creative, is what we can use to help us in inviting communities to the green table.

Mixed Messages to Adults

As adults, many messages we receive about play are negative:

How childish.

It's just a stupid game.

That's child's play. (Meaning it's really easy, and so you "should" be able to do it.)

On the other hand is "All work and no play make Jack a dull boy." So there is some acknowledgment that play has value.

There's a lot of judgment about play being "too much" in certain situations. You're not supposed to be too playful at work. Put on your serious face, and just work, work, work. But what if that attitude is actually robbing us of some of our best and most useful work?

Getting Us into Space

Let's check out a real-world example of play at work, in this example highlighted from Stuart Brown's book.

The Jet Propulsion Laboratory at Cal Tech is where you go if you want to make a splash in the space program. They were the brains and engineers behind the space shuttles, the Mars Rover, and more. They are the go-to, solve-the-tricky-problem, get-it-launched-into-space people. But they started to run into a big problem. Their engineers, while bright as could be, with oh-so-shiny grades from the fanciest schools, were not solving the challenges sent to them as well as previous engineers at the company had done. Despite their paper smarts, these current folks were not creating practical, workable solutions for the tricky engineering challenges faced by the space program.

So the Jet Propulsion folks looked back in their history to look at who had done the best at making inventions that worked. They spotlighted folks who were members of the teams that engineered and created their best inventions. They asked themselves the question, "What did those successful engineers have in common?"

It turns out that it was how they played when they were kids that helped to turn those people into successful engineers. Those folks were the ones who took apart radios and rebuilt them, created their own derby cars, took apart all kinds of stuff and put it back together differently—just to see if they could, and to see what they could make that was new.

So now Cal Tech has a new hiring policy. They ask, "What did you do as a kid? What were the things you enjoyed? What did you play with?" The engineers who had been "Let's take it apart and build a new one" kids become the new, can-do engineers of the Jet Propulsion Laboratory. How could you look at play differently in your green organization? What's been missing from your office, hiring policies, events, or how you serve your community that you could add back in?

Benefits of Play

Let's take a look at some of the other benefits of play. The good news is that the benefits of playing are multiple and generous. Things that are *so* much fun to do are good for us and help us in a variety of ways.

Physical. When we play physically—run and jump and play like we did as kids—our bodies benefit from the exercise and fresh air. We literally bring more oxygen into our cells. The benefits to exercise are huge—reduced stress, better sleep, a stronger and more flexible body, and so much more.

Mental. Stretching ourselves mentally—with a tricky crossword puzzle or a new game—gives us new ways to solve problems, or to combine things in new ways. Think of an excellent cook you know, and consider how s/he combines new spices, ingredients or textures into the food s/he makes. That willingness to try new things is another way to play.

Emotional. When we bring a sense of laughter and play into our relationships, that can smooth the sometimes bumpy ride that is relating. We gain a greater sense of trust in the relationship, that it has some *give*, like a rubber band—an ability to stretch and change over time. This resiliency is something to cultivate in yourself, and with your volunteers, colleagues and in collaborations and partnerships with other agencies.

Community. Play creates community, for example, at parks and picnic areas, playgrounds for kids, venues for community theatre, a national park where families and friends can enjoy the splendor of nature. Play creates resources—the best-selling spy thriller novel, the latest kitchen gadgets, the Internet, new medicines. Play even creates profits. Look at how many people buy tickets and attend live sporting events, pay for a snack and a program while there, and are thrilled to be there and be a part of the game. Theatre productions and musicals, engaging books and movies—all are play. Community celebrations and dances are created through play, on scales small and large—

from me and my kids dancing in our living room to Mardi Gras in New Orleans.

In communities, the benefits of play mean we can create an interconnected web of people who can be creative and solve problems together, even about something tricky, like disputes about water rights, or downstream pollution of a key fishery. Play creates more trust, more connections, and more happiness.

Take a moment to consider just how much we have invented, how much we have created, with our marvelous, flexible brains.

Rules of the Game

Play has rules and guidelines. In the realm of sports, each game clearly defines what is play, and what is not play. Terms like "out of bounds" and "foul," and flags for violations describe that which is not play. There are specific explanations for how to play the game, such as how many players are included, where the game is played, what the rules of play are, and the exceptions to the rules. We need these to help us define the game.

We can see and understand this principle when companies or individuals pollute groundwater, allow the erosion of our land, and fill the air with toxins, because these actions are outside of the rules, a violation of the rules of fair play.

Here, at the rules of play, my friends, is where the rubber meets the road. Learning the play rules for different communities is how to best and most politely know how to engage with them and how to invite them in as full participants. This is something you'll need to research and to discover in your own community.

As you know, there are "rules" that have been twisted from the get-go; rules that simply are not fair from the start—like all of the "-isms" that say, "Oh, you can't play this game because you're Latino" or "...you use a wheelchair" or "...your family doesn't have enough money." These "rules" are discrimination, plain and simple. They are bull. Malarkey. Crapola. These are the ones for us to eliminate from our own minds and from our communities. Let's call forth and *create* rules that are fair, just, and honorable. They are also way more fun!

Community Celebrations

You may already have a community in mind for your green organization that you would like to engage. One of the best ways to find out how to interact with them is to find out what they do when they celebrate. Consider the ways they gather already, and clarify the answers to these questions:

1. What does that community do when celebrating? (There will probably be lots of ways throughout the year. I suggest picking a couple that feel especially welcoming and appealing to you, and that are open to folks outside of that community.)

2. Are there dress codes? Is there special ornamentation that is sacred and only wearable by certain members of the community?

3. What happens at this celebration? Are there certain kinds of activities that always happen, such as dancing, sharing food, storytelling, or special stories told via theatre?

4. What are the codes of conduct? What's considered polite or impolite? (This may require some thoughtful and considerate research, and may require your willingness to be open-minded.)

5. In that community, there will likely be interplay between the traditional types, those wanting to keep things the same, and the ones creating something new. How can you politely invite people to join you? Who seem to be the open-minded ones of this group, the ones who are respected for their ideas, even if they are new or different?

6. Who are natural bridges into that community? There will be community leaders, teachers, people who volunteer in their own communities and others. These folks care about their people, and their welfare. They could be your natural allies.

This information will show you how that community of people likes to gather, and how they celebrate. You'll discover more about what they, as a culture, prefer, and what's outside of their comfort zone. Maybe this community, as a whole, loves music, and so that is an integral part of their gatherings. Another community may prefer storytelling through elaborate theater and special clothing.

Gathering and considering this information carefully will help you create events and ways for them to participate with your green agency that feel welcoming to them.

Sprouts: Your Green Take-Home Tips

1. What community would you love to invite to participate with you and your green organization?
2. Find out what they love to do when they gather, and find out what basic polite behavior is to them.
3. Invite them to participate, and invite them to be part of the planning process. See if they can be the ones to plan and lead the event, and you host the space for them.
4. Figure out a way to integrate that community into your next event, and be right alongside them.

Renewing your relationships with people helps to renew our care for the earth. The following chapter explores how to use play and connection to build and renew your relationships.

Chapter 6
Renewable Relationships

"Laughter is the closest distance between two people."
–Victor Borge

H ave some of your friendships or working partnerships become dull or stale? Do your meetings feel like eating old, dry toast? Yuck. This chapter will help you renew old relationships and establish new ones.

We all need friendships and relationships with people outside of work. We humans are such social creatures; we thrive on healthy community and interrelationships. Healthy relationships help us to honor the relationship between ourselves and the Earth, to better care for the land, water, and air, and to enjoy the many interrelationships between the trees and the wind, the land and the animals, and so much more.

Playful Relationships

So, with an eye on your current relationships, let's take a close look at the definition of a playful relationship.

The following definitions are from Dictionary.com:

Playful (adjective)
1. full of play or fun; sportive; frolicsome.
2. pleasantly humorous or jesting: a playful remark.

Relationship (noun)
1. a connection, association, or involvement.
2. connection between persons by blood or marriage.
3. an emotional or other connection between people: the relationship between teachers and students.

Do these definitions describe your relationships? Do you have fun connections with others? I'm talking about your close friendships and best buddies. How about with your partner, husband, wife, or sweetie? Think about those friends from college that you have a picnic with every year. And then there are the wonderful folks you can call in the middle of the night if you're sad or stuck. If you're in a situation where you feel like you'd like more friends, or closer friends, stick tight, I'll address that shortly.

Resiliency

When we have resilient relationships, we have created a rubber-band-style elasticity to our relationships. There is a give, a flow, a sense of movement to the relationship that allows us to stretch and move within them and be able to roll with conflicts or difficulties that may arise.

Rather than settling for a neglected rubber band that has become dry, fragile, and easy to break, let's keep our relationships pliable and stretchy.

Adventuring

Pretend I've given you an entire day and evening to go and play, simply to have fun, to have an adventure. You've met your friend or partner, or your group of friends, and you have all of this time to enjoy each other's company in a place you love.

Maybe you're going hiking in the mountains and plan to camp overnight. Or perhaps you've planned a day with a beautiful lunch and a walk through gardens. Maybe you've planned it out carefully ahead of time, and have relished the anticipation. Or maybe you and your buddies are more spontaneous types and tend to get together and do things on the fly.

Imagine that play time specifically—the sights and sounds, the food and jokes. Soak this in—the laughter, the beauty of the setting, and the joy of being together. It's deeply fulfilling.

Now travel back to the here and now.

Take some notes about how that day went.

- Where did you go? Someplace familiar or unexpected?
- What did you do?
- Was there something wonderful and unexpected that happened?
- What was particularly enjoyable to you?

After this imaginary outing, you have a pretty good idea—or maybe even several good ideas—about what you'd love to do with your friend or friends, about how you'd like to reconnect and have fun together.

Renew or Deepen Current Relationships

Sometimes we get distanced from friends because of illness, or work responsibilities and, pretty soon, it's been a while since we've seen or talked with them. Here are some strategies to help renew that relationship.

Ask for Playful Time Together

I invite you to say to your friend, "I'd like to add more playfulness to our relationship; More joy and laughter, silly times and fun activities. I'd like to have a play date once a month with you, where we take turns planning something fun for us to do. Of course we'll still be there for each other when things get rough; I just want to have more playtime together."

And you can negotiate from there. Maybe there's a deep play deficit in your relationship, and it'll take time to fill it. Start with silly video clips, and graduate to funny movies. Try out different genres of play.

Tea for Two: Identifying Play

Getting together with friends is such fun. We want our friends to have just as much fun as we do. How do you know if it'll be fun for each of you? Here's how you know what play looks like for both of you:

- It is not work.
- You both agree that it's play.
- You both feel rejuvenated afterwards.
- Both of you would love to do something like that again.

Obviously this may require some willingness to experiment. For example, you're both like, "Yes, let's try foosball," and one of you likes it, the other, not so much. So you try something else and discover that you both love Ping-Pong© . Bing! Bing! Bing! Alert the media! That is play for *both* of you.

David, my sweetie, loves the band Led Zeppelin. Me? Not so much, or so often. So he plays it when I'm out, or plays it with our son, who also likes it. That way, they can play the way they want to, and when we want to play together, we do something all three of us enjoy.

Set up a Recurring Play Date

Repeatable play is something you can put into your calendar so you can keep the cycle easy and sustainable.

Pick a date, like a monthly or weekly day, when you and your friends can gather together. Dates early in the month are usually easier to remember. For example, set it up so that you pick the first Saturday of the month, and you meet up. Even if you cannot make it one time, you'll know that you will have the chance again the following month.

Never done this before? Or feeling a deficit of friends, too? That's okay. It can be organic. Just ask two people you really like (who are not into drama or being mean) and invite them to invite two people they really like to a monthly gathering. This is an opportunity for you to invite community and diversity into your own life. (Hint, hint.) Meeting every month allows you to build relationships over time, and the security of knowing you'll meet monthly allows for the group to relax together, knowing you'll be showing up and continuing to show up. You'll soon have a group of people who like each other and enjoy getting together. Trust and connections will build.

Renewing a Friendship

If you feel like you have not hung out together and had fun enough with a particular friend for a while, someone whose company you simply enjoy, you may be experiencing a play deficit. There are some simple ways you can change this:

Teeny, Tiny, Silly Steps

If it's an old friend, see what you can do to encourage playtime together. Can you meet for a phone play date? Just to call her and gab and tell each other silly stories? Or text a funny picture to him. Take a small step in a playful direction, and then keep building on it. This should feel easy-peasy to do, like you could do it while half-asleep.

These little steps may not feel like "enough," but notice when, put all together, teeny, tiny steps will take you wherever you want to go. And, as a bonus, they lighten your day and your mood, and strengthen your friendship.

Playing off the Beaten Path

Do something new! Something "crazy," like karaoke. Or Ping-Pong©, or that new computer game your nephew goes on and on about.

Maybe you and your friend are more of the prim and proper types, and normally have tea together, but both of you want to try being all-out adventurous and go horseback riding or snorkeling or kayaking.

Giant Steps

Leap in! Plan something new and wonderful! Plan a friendship weekend getaway. Each of you can bring your favorite books, your personal playlist, your best stories and memories of each other.

Or create an active adventure together. Travel to someplace new together, or have a "staycation" at home and use that time for connection and conversation, and doing things you both love to do together.

We need our environmental peeps to be happy and fulfilled. We need you and your buddies to be refreshed, to be able to do the work to change the world.

Creative Conflict

Sometimes, even in the best of relationships, there is conflict. Let's discover a couple of ways to handle it that are practical and fair.

Fight Right Rules

Fight right! These rules to fighting right are from *The Happiness Project*, by Gretchen Rubin. They can help you and your partner or friend or colleague keep a conflict to one area, and they can actually forward your progress through the conflict to the other side.

Fight right rules:

- Stick to only one topic at a time.
- Be specific.

- Contain the amount of time.
- Agree to disagree.

Your Friend the Egg Timer

Little known fact: egg timers love to help solve conflicts. These simple devices can help you, too. My husband and I used this method when we were dating, to help us solve conflicts. It's so simple, and so helpful.

1. Find an egg timer, or use a smartphone timer or other handy timer.
2. Set it for five minutes. The first person to speak gets to talk for five minutes without interruption. The second person listens, and is present, for this.
3. Set the timer for a new five minutes. The second person now gets to talk for five minutes. The first person listens attentively, and is present.

4. Repeat this sequence, if needed, so you get all of the basic facts out.

Sometimes you'll find that the conflict is a simple and basic misunderstanding that gets handled in the first set of five-minute intervals. Other conflicts may take more time. The timer allows for a neutral container that is to be respected.

Additional rules include:

• Listen attentively.
• No yelling or raised voices.
• Do not use the time to construct a rebuttal; you'll miss key information that may dissolve the conflict quickly. Listen to *understand*; do not use the time to construct a reply.
• *Helpful Hint*: This is not speech class or a time to worry about sentence construction. The two of you may speak awkwardly. The point is to be authentic with each other.

In your five minutes, you are to aim for clarification, and to say things about yourself, like, "This hurts and does not work for me. You can say things about the other person like, "I would prefer if you said it this way…" or "This way works best for me." Another handy phrase is "I'd like to find a compromise that works for both of us."

If the timer dings and you're in the middle of saying something, finish your sentence, then pass it over to the other person. This is about talking it out, and finding out more relevant information, so together you can find a solution that works well for both of you.

Wear a Hat

This tip is from Bill O'Hanlan, a certified professional counselor. It's so practical, and it's even fun. When arguing, especially if it's an argument that you have had over and over again with someone, and you're stuck, and you keep repeating the *same* darn argument—do one thing differently.

You can have the same argument, but you have to wear a cowboy hat.

You can have the same argument, but you must spin in a circle or hop on one leg while you do it.

Or, while both of you are clothed, sit on the toilet while your partner sits in the bathtub. Somehow, the ridiculousness of sitting on or in porcelain while arguing changes your perspective, and may have you giggling.

This will help boost your brain into new territories, help you loosen this old pattern a bit, and get to a place where you can see things a bit differently. It may also have you both laughing at yourselves, and breaking up that old pattern even more, which is both fun and helpful. See if you, together, can take a next step as a result that feels new, or feels like a fresh pattern.

Meeting New Kindred Spirits

Kindred spirits are those kind souls who you have an immediate kinship with, whatever race, ethnicity, religion, age, or ability they may be. Notice them. These are your people. These are the ones who will "get" you, despite and because of your fumblings. When you meet them, there may be a feeling of a *click* or a deeply felt *yes* or even a jubilant "Hooray, hooray, I've found you!"

They are dear to you, and you root for each other. Open your eyes to the possibilities of finding these people, your people. Start with the people you already interact with. Or, perhaps, when I say these are the people who "get" you, you already know exactly who I mean. It may be someone from long ago, who you think of fondly and hope to rekindle a friendship with. Or it may be someone entirely new.

Your next kindred spirit might just be:

- The thoughtful librarian who helped you to find just the right book.
- The kind woman you set your mat down next to at Pilates class.
- The smiling, relaxed man you keep seeing on your daily walk, or at the co-op, or at the grocery store, and you've wanted to say hi. Say hi this time.

Maybe, as you're reading this, there's someone coming to mind, someone you'd really like to know. Look through your list of contacts, or the folks you know through social media, and your Facebook friends. Is there anyone who springs to mind? Anyone who you'd like to know better?

Or if you're an introvert, see if your extroverted friend will introduce you to someone they think would "get" you. (I'm an extrovert; I love to help with this kind of introduction.) People are willing to help, if you just let them.

You can always start by smiling at them.

Safety Rules for Grown-Up Playdates

Sometimes it helps just to have some basic rules in mind about doing things together, whether you're meeting old friends or new. And, yes, if you'd like, you can say these out loud to each other. You can introduce the topic by saying, "I'd like to operate with these guidelines in mind. Do they work for you? Would you like to add anything?"

Then share these Basic Rules for Playdates for Grown-ups:

1. Be respectful.
2. Be kind. This one is key!
3. Find out what play is for that person; know what it is for you, then find the points that match, and combine.
4. Let's agree that if we have a conflict, we'll resolve it respectfully and patiently.
5. Check in—Is this working for you? For me? What are we/ are you/ am I loving about this? What could we adjust to make it even better?
6. Check out—How did the event/lunch/etc. go for you?
7. What would you like to do next time?

Some of this you can tell by simply observing. Is the person laughing? Talking, and interacting? Is there a relaxed vibe, or an engaged vibe? It's okay if there are awkward moments; that is fine. There will be, especially if you're just getting to know each other, and even after that.

Here's a basic safety caveat: If you find yourself feeling like you're near a big, growling dog with teeth bared, except that this is a human, get yourself quickly and safely out of there. Trust that gut of yours. It's a good one; it works. Said another way, if there's any time when the hair on the back of your neck is going up and your body is all tense and saying, "Alert, alert! This person is *not* safe," then get away. Friendships are for having with safe people, kind people, people you feel safe around, people who treat you *really* well. And I'm not talking about the kind of uncomfortable you'll feel when you're challenging your own biases, or the biases of your organization, or when you are stretching yourself, I'm talking about basic safety and respect here.

And back to our conversation. Enthusiasm will carry you places where politeness dares not. When in doubt, be curious. You can say things like:

- Please say more about that.
- I'd love to hear what you have to say.
- I'm fascinated. Please continue.

Sometimes, a companionable silence works the best to allow a quiet person to complete his or her thoughts, or to simply allow space around new ideas or ways of relating with each other.

Sprouts: Your Green Take-Home Tips

Practice having healthy, resilient relationships in which you feel refreshed and inspired.

1. Resilient relationships help build resilient environmentalists.
2. Have fun with your buddies! (That's the essence of this chapter.)

Being inspired and connected will help you in your work to help others have a healthy relationship with the Earth.

Now you've learned some ways to renew your current relationships. Next we'll dive into your plan to invite diverse and wonderful communities into your organization.

Chapter 7
Making Your Plan

"My top priority is for people to understand that they have the power to change things themselves."
—**Aung San Suu Kyi**, Democracy Activist

Now that you've had some inspiration, ideas, and created forward momentum about your wish to create a more diverse environmental organization and community, it's time to move those ideas forward into the world.

Planning for Successful Fun

Let's make a plan. Get out your favorite markers and a big sheet of poster board, or your wonderful whiteboard (could be an electronic whiteboard) and get ready. Get ready to have fun. How are you going to make this fun for yourself? Use your favorite tools from this book. Play fun music. Watch a silly puppy video or crack a corny joke with a friend to get yourself into your playful and creative space.

Put on Your Thinking Cap
Below are some basic questions that need to be answered as you make plans to invite diverse communities of people to the green table. These are simple questions, but that does not necessarily mean that they are easy to answer. Answer them thoughtfully and well, and you'll be on your way to creating a plan that works.

- *Who* would like to join your movement?
- *What* would you like to invite them to do? What would they like to do?
- *When* would be the best time for this?
- *Where* would be a place where this community would feel welcome?
- *Why* would you like to invite this community to join you?
- *How* can you create this together?

Start with one group of people, and finish one set of these questions per group. This will help you focus and create a plan that works for that specific group. If you'd like to, later you can compare the similarities between groups, and choose activities that many or most of them would like.

Write It Down, Draw It Out

As you answer and brainstorm about the questions, type or write quickly to get the basic answers down. Set a timer, if need be. You may not know all of the answers to some of these questions, and that's okay. There will probably be some research you'll need to do. Plan to go back and fill in and expand your answers later. For now, we want a quick sketch to help you get started.

If you find yourself getting stuck, write down your first response and then move on. You'll be able to avoid your inner critic this way. If you respond better to verbal questions, have a friend or colleague ask these of you or of your group, or make a simple recording for yourself asking the questions, then play it back and answer those questions.

Take about ten to fifteen minutes for this first iteration of answering the questions. Jot things however they come out. If just phrases work better for you, that's great. This does not have to be sentence-structure perfect. Or maybe you like to doodle, or create word-pictures as your form of notes.

Now you have a basic plan to play with. This is your first idea, and it will help you get set up for something new that could really work for you.

This initial process may seem simple, but you can create an event that works really well by having good answers to those basic questions. We'll dive more deeply into these basic questions in a bit. First let's flesh out a few of the logistics.

A Variety of Events

There's such a wide variety of events that you and your community could create together. Here's a sampling of some of the different types of events.

Themed Events

These are events that have a specific theme, like "Racial Equality and Eco-Justice for All", or "Clean up the Local Playground" or "Solving the Water Crisis in Milwaukee" or "Celebrating Earth Day." They may be planned far in advance and be something you'd like to arrange speakers for and decide on a set location, invitations, and a presence on social media. Or they might be more casual events that focus on welcoming and celebrating a new community to the green table.

Combination Events

These may be events that fit together, like an afternoon about "Social Justice in the Media" followed by "Buy Local—Support Your Neighbors & Save the Planet."

Basic Meetings

Regular meetings do not have to be boring; on the contrary, they can be interesting and create intersections, connections, and real change. Re-think the meeting! Dull and boring meetings produce dull and boring results for your organization. Creative and diverse meetings generate creative and diverse results.

Your last Board Meeting was so boring that you wanted to rename it the Bored Meeting. Yawn-a-rama. So, what can you and your people do to invent something different for your next meeting? We've talked about bringing in music, including different kinds of play, and learning from the animals. What inspirations have you had that you can use in your next meeting to liven it up and bring in more playfulness and innovation?

Pop-Up Events

These events address a certain need and happen more spontaneously. They have a time-sensitive aspect to them. For example, people may want to protest a proposal for a polluting coal-fired power plant to be located near a local church, synagogue, or sacred site.

Seasonal Events

Seasonal events may relate to the calendar of the natural world in your area. For example, you could do a "Tree-Planting to Honor Our Elders" week or host an "Eco-Justice Harvest Festival."

Think about the seasons and seasonal events in your community. How could you create an event that uplifts and celebrates them and has natural ties into your green agency? For example, harvest time would be a great time to have a volunteer appreciation party. It's a season of abundance and gratitude. Spring would be a good time for new beginnings and inviting new connections. Many cultures recognize and celebrate the seasons; it could feel natural to include them and plan an event together.

Recurring Events

You're creating something new and fresh and wonderful that you envision happening again next year, or next month. Or maybe you're reviving something that you did before, something that people loved, but you have a new angle on it, and new groups of people will be participating. This is cool. My bet is that people will want to do it again.

Sketch out in your mind when the next event will be and how it could be a recurring event. It could be seasonal, or tied to another event, or simply a monthly meeting that you've already been thinking about. What could be a natural follow-up to this event?

If you're new to this type of planning and can hardly imagine life beyond the next event you have planned (I call this *event-itis*), ask a friend or colleague who's not so deeply involved to help you brainstorm what might come next, or at least the approximate date when your next event would be.

If you have brainstormed some ideas in community, as I hope you have by now, you may have a whole list of things to choose from.

When people attend your current event, they'll naturally ask when the next one will be, and you'll want to have an answer for them. Be specific, if at all possible, even if you do not have an exact date nailed down yet. You can say something like, "We're planning something for September or October, and we'll let you know via email. We'll share an announcement to your Facebook page, and an announcement at your group's meeting."

Community Savvy

Where do you begin? How do you know you're going in the right direction? Remember to relax, and remember that human beings are naturally good at connecting. We are a social species. Here are some ideas to help you get started with planning in more detail.

Mutual Respect

You're excited to connect, but terrified of making a social faux pas that insults someone. So, first, check your intentions. Are they clear? Do you want to connect from a place of service, a place of good-heartedness? That's the best place to start—with a clear heart and good intentions. You'll likely make mistakes, and that's okay. We all do. That's what apologies and fresh starts, redesigning and trying again are designed to handle. Most people will recognize your good intentions, and have flexibility in their response to you.

Coffee and Community

Now, what can you find out ahead of time, once you've decided on an event? Invite someone savvy about the community you'll be including to coffee, or lunch, or simply meet to talk. You might look first at people you know, or at people they know. Be specific with your question when you put out feelers. For example, you can say, "I'd like to know more about how I can be a more polite person, and a better friend and ally to the Black community. Do you know someone who would be willing to chat with me? Can you introduce us?"

Listen

When you meet, do your best to listen more than you speak, and ask follow-up questions in response to what the person says. Be sure to thank them for their time and attention. Tell them that you are a beginner in your interactions with their community (if you are), and that you'd appreciate basic tips for being polite. Pretend that you are an ambassador to the United Nations.

When you meet your contact, here are a few questions you could ask:

- What's something I could do that would welcome people? How could our organization be welcoming?
- What's something I should watch out for? What should I not say, do, or imply?
- How can I be polite, friendly, and welcoming to your community?

Rule-Breaking and Uncharted Waters

As you move forward with planning your event, you may break some unwritten rules, and that is okay. Certain rules need to be broken, especially ones that exclude people. Stomp those into the ground!

Branches of the Community Tree

There may be groups within a community who are not always included; this will vary widely. I'm mentioning these folks in particular to lift up their value and capacities, and to ask you to remember to include them in your planning and to invite their multi-layered contributions.

Little Ones, Teens, People with Disabilities, and Elders
Including these groups will depend on the type of event and involvement you are looking for, but I strongly suggest having at least a few events per year, or every six months, that is friendly to families, where babies, teens, people with disabilities and elders are welcome.

Sometimes we overlook these key people in our communities, and when we create opportunities for them to participate, we also gain. We gain the clear-eyed innocence of the young child, the needful rule-breaking of the teen, new ways of understanding and doing things from folks with disabilities, and wisdom, patience, and deep, important questions from the elders. Multi-generational participants, and people with differing capacities, have much knowledge and creativity to share, and many will be glad to be asked.

Open Time and Open Space
Sometimes people need to process and move forward within swathes of time that are open and unstructured, to be able to contemplate, and to grow what's needed inside. As a formula it might look like this: *Open time + Open space = Open minds.*

Many cultures are not as ruled by the clock as we are in the Western world. We would do well to emulate that stance, to allow more free, unstructured time in our days, in our meetings, and in our time together.

Make space for new ideas and connections. Your internal environment can reflect the external way you organize things, and vice-versa. Hierarchical events tend to host the speakers in the front. Events with more of a community focus, and an emphasis on contribution from everyone, tend to seat or organize people in circular or semi-circular settings.

Consider this carefully as you design your event. If you can, move chairs and seating areas around so there is literally an open invitation to participate.

Inviting Contribution

People want to feel needed, to feel valued. They want their thoughts, ideas, and opinions to be respected. Give them some way to give back, to be involved. We are a highly social species, so most humans like to be together. This is especially important for the do-ers of the group, the ones who like to contribute by doing something. Give them something to do. Small actions count. They add up to bigger actions.

Consider mixing local events with regional, national or international speakers and new ideas through books and other media. It's a great way for people to act locally and think globally. An excellent example of this is the Bioneers Conference. It's a yearly gathering in California that honors "people, ideas and actions that merge human creativity and nature's genius." In 2016, the Bioneers hosted their 27th conference. There are regional Bioneers' events around the country that host local and regional speakers. There's a macrocosm and a microcosm here, and joining them combines people and ideas, which means more opportunities for creative connections and new ideas.

Diving a Bit Deeper

Now that you've had a chance to do some research, and to contemplate a bit, let's look back at the basic questions you'll need to answer to invite new people to join you and create a gathering.

Imagine that you're creating a party, one that'll be fun and interesting for people who attend. In fact, go ahead and create a party, and sandwich meetings into the party. Think of the event as a layer cake, with layers of cake (meetings and work time) and frosting (games) and fruit (music and wordplay).

Let's explore your next steps for organizing the event you've decided to put on.

Who?

Who would like to join your movement? Who would you like to have join you? Who is missing from the picture? Who would be a wonderful and creative partner?

You might be looking around and noticing that your green group is not very diverse. Maybe you're not sure who is even in your community. You may have just moved there and are new to the area. You can check online to find data from the U.S. Census to get a general overview of the different ethnic groups who live in your area.

Other ways to find new green peeps include:

- Checking to see if there's a senior center or a retirement community.
- Asking at the local Chamber of Commerce for a list of groups who might be interested in your green work—like birders, hikers, canoeists, and more.
- Looking at the bulletin boards at your local library; often, new groups will post invitations to come to events there.
- Checking on social media to find folks, via Meetups, Facebook, Twitter, etc. (Keep in mind that these folks will be the tech-savvy contingent, so they may not include everyone you'd like to invite.)

What?

What would you like to invite them to do? What would they like to do?

Hopefully, you've started this process by talking with someone over coffee, or had a chat (maybe online, via Skype, Zoom or FaceTime) and then met with members of this community several times, and your new group has some ideas.

Exploring the *what* is where partnership and true listening is key. Maybe there's an obvious challenge that you know about for that community already, and you have some ideas about what could help. Do yourself a favor, and do them the courtesy, of asking what *they* would like to do. They'll know much more intimately than you what their community needs.

It is respectful to first ask, and partner with them, then work together. Do not be a group that comes in, takes ideas, and then leaves. That's rude and annoying at one end, and at the other end it could be disheartening to the whole community, or even cultural appropriation.

Ask what they would love to improve on behalf of the community? For example, better access to healthy food, putting a school garden in at the local middle school, starting a new farmer's market on the east side of town, creating a food truck to bring food to marginalized communities, having easier access to public transportation.

Ask what's been a thorn in the side of the community. What's been a long-standing problem or challenge? How could you work together to improve that situation?

When?

When would be an ideal time for doing this event? Find out if this community has a special celebration where you could join in. Conversely, are these folks really busy at certain times of the year with special celebrations, and so would not be available to join in a co-created event then? This will be a dance to see what works for both of you.

Where?

Where would be a place where this community would feel welcome?

Is there a place where this community already meets, that is theirs? Or ask how you could make a space that's known more friendly to them.

Consider natural spaces—under a sacred tree, at picnic tables by the lake, at a local park near the kids' playground. You could rent an outdoor shelter for the day, invite people to a potluck, and have your meeting there.

Why?

Why would you like to invite this community? Why should they work with your group? Is your organization best situated to serve them?

This is about the big picture. The simple answer to these questions may be because they have been underrepresented, or not represented at all, in our green organizations. This is unfair, and maybe you'd like to change that.

Exploring the *why* may require some introspection, deep honesty, and maybe even a smidge of humility. *Can I serve this community? Are we the best group to partner with them?*

You'll need to leaven this exploration with some courage, however, because it's easy to say, "Oh, this will never work," and not even try. Look at the reasons behind why you want to partner with a certain group, and explore whether or not it will also serve that group of people. We are looking to create a long-term, win-win partnership here.

How?

How can you help create this? Can you partner with others? How can you make this happen? These are the practical ways you'll connect and work together. There's a toolbox full of ways to do this, and you likely already have a toolbox with some tools to use in this situation. Some tools will deal with technology, others with real time savvy.

There are all kinds of ways to connect via technology—on social media, like on Facebook, Instagram, and Twitter; there's email; there are meet-ups and more being invented by the moment. Choose the ones that work for you *and* for the group you are inviting.

If there is a technology gap, some people will have technological capacities and others will not. People who are elders or are in marginalized communities may not have easy access to technology, and you'll still want to include them. So you'll need to adjust—and send paper newsletters or use other means to communicate with them. The personal touch, in these days of highly technical communication, can work wonders. Sometimes it's the personal invitation that's hand-delivered, the phone call, the hand-written note, which people most notice and appreciate.

Completion and Celebration

Together with your community, it will be helpful to create a way of knowing when your tasks are complete. In creating a working eco-community, there will be milestones along the way. Note those so that you know that they have been completed. Otherwise, you may discover that you've done what I call "moving the goal posts," changing the goals before allowing yourself to actually finish a project, or to feel like you've finished. In so doing, you could get stuck and make more work for yourself than is necessary, and that's not helpful.

Allow yourself and your group to have the satisfaction and the joy of actually finishing a project together. Celebrate not only at the event, but that you pulled it off!

You can celebrate as you go, and it's helpful to clarify in advance what the key milestones along the way are, so you know your milestones, and can celebrate them.

Milestones

Milestones are special achievements made by your organization. Some of them may signify minds that have been changed, or new participants gathered, or new communities formed.

Here are some examples of milestones you could recognize:

- Having a new board member who represents a community of indigenous people you had not had represented previously.
- Your big spring event had sign language interpreters for several attendees who were deaf, and this service was enthusiastically received.
- You met at an accessible community center for the first time, and it was much appreciated by your volunteers who are elders, and by a new staff member who uses a wheelchair. It worked beautifully for your entire group, and will allow you to partner with an elders group in the near future.

Celebrate Your Accomplishments

Celebrate what you have done together. This allows people to have a true sense of accomplishment, of the joy of doing something that may have been difficult, or that required patience and many steps to be successful.

The simple practice of gratitude is key here. Your thanks and heartfelt appreciation for what people have accomplished will feed their hearts. Being noticed and appreciated is something people crave.

Acknowledgment can be as simple as a certificate mailed out to your volunteers, or as elaborate as giving a fancy Volunteer Appreciation Party. Perhaps you ask your board members to call special volunteers or active citizens to thank them for all of their good work on behalf of the community.

In my years as a volunteer coordinator, I would visit the volunteer sites, and thank people that very day. I also noticed that when I called people up again the next day to thank them personally after their day of volunteering, that was so appreciated. If it's a particularly large group of volunteers, you may most easily do this for your key volunteers, or your lead volunteers, and send out thanks in other ways to your bigger bunch of volunteers. If it's an event where people are volunteering, you can circulate throughout the day, giving your thanks for the participants directly to them, in person. People so appreciate being appreciated. I've sent out an e-newsletter after the event, with pictures from the event to people, too.

Be sure to keep a list of who has done what, and what has been donated, so you can thank your generous donors for their time and material donations.

Beware the Sneaky Biases

Here's a heads-up. During this process, you may notice your internal biases showing up. You may feel reluctant to call someone who it makes perfect sense to call. You may notice that certain images from movies or TV make you uncomfortable. These internal biases can show up in a number of ways. This is okay. It's a sign of growth for you. Do some more thought work. Also, once you've made a trusted friend in that community, you may want to bring up this issue, when it feels like good timing. You can say something like, "I've noticed some things that are bothering me. I was wondering if this is a stereotype." Then say what they are and ask, "What do you think about this?" You may be able to have a rich and respectful dialog that will strengthen your trust in each other and in the work you are doing together.

Fresh Leadership Patterns

We need different inputs into our green organizations so we can get different outputs. Sharing the leadership leads to new ideas and new ways for people to contribute. Even conflict, when attended to well, can be creative, because different ideas coming together and bouncing off of each other can create new ones. Think of Ping-Pong© balls bouncing off of each other into new trajectories—generating new ideas works the same way.

The capacity to share leadership and to collaborate is essential for welcoming new and diverse people to the green table. If you have only used a hierarchical model, where one person is the leader and there are levels of leadership "under" that person, I invite you to try others.

There are folks who dive deeply into the theory and practice of these models. Here are some quick and practical pointers on a few of those models.

Circle 'Round

Leadership can travel around a group. Each person gets a turn to speak. Sometimes, a sacred or significant object is passed from person to person to signify that one person has completed their turn and another's turn is beginning. Everyone is responsible for listening respectfully and attentively. This means actually hearing what the person is saying, rather than formulating a response in your head while they are talking.

When starting, it can be useful to have a timer as a neutral boundary-keeper, to help with this. (This could be our favorite, the egg-timer, or a timer on someone's phone, for example.) In this way, each person in the circle gets an equal share of attention.

As the group grows in their capacity, and builds up practice with this method, you can simply pass a special object from person to person, and each person can speak at their own pace, without the timer. In certain Native American traditions, the object passed around was called a "talking stick," and the person with the stick was the one whose turn it was to speak.

People in the Wayland B house, an international, interfaith cooperative where I lived for almost two years as an undergrad, practiced shared leadership. We took turns leading our monthly meetings, which allowed people to practice their leadership skills. Quieter people also had their turn to lead us, as did people who had English as a second language. It was an excellent experience in learning that different modes of leadership work just fine. Leadership does not have to look like one charismatic leader directing things from the front of the room. At the Wayland B house, we met around our dining room table, had great discussions, did lots of problem-solving with much laughter, and respect, and often shared a meal together afterward.

Square Up or Semi-circle

In this leadership option, two or three people co-lead the group. They take turns with presenting material, asking questions, and inviting participation from the group. This is a useful format when a small team of people has been in charge of leading a project, and a couple of members of the team can present to the larger group.

The Point

One person leads the group. You're probably already really familiar with this model. One person leads and directs the group, the others listen.

This can be useful if one person has a specialty, or unique experiences to share with the group. This model may work well for guest speakers, trainers, and experts. Even so, you can invite your guest speakers to create another, more interactive model for their presentation.

Patterns in Nature

We can also look to the variety of models that nature presents to us for solving our human difficulties.

Biomimicry is about looking at patterns in nature to see how something is accomplished, and using a similar technique to solve our problem. In nature, a tree solves the problem of getting water from the roots up to the far-most leaves through a series of repeating flows, via capillary action. How might this concept be adapted and used to create more diversity in our organizations?

We think of water flowing down, as in waterfalls, or out of our faucets. But water flows *up* in a tree! Capillary action involves tiny cells in the tree's roots taking in water, and then passing it to the next cell up, which moves it into the next cell up, and so on. (With apologies to the plant scientists who may be shaking their heads at my highly abbreviated explanation.)

This action is natural to the tree. It's easy, and part of its "routine." Capillary action is not separate from the tree. Different cells of the tree are working together to support the whole.

How could we invite more people in and help them stay as "part of the tree" in our organization? How could we make diverse communities be simply part of our organization, not seen as something separate. How can we make them integral?

Thanks to new research, we've discovered that the trees "talk" to each other. Old logs are nurtured via a network of roots. Communication happens through tiny rootlets and within the soil. Different kinds of trees, living close to each other, communicate via a sophisticated, interwoven network.

What kinds of messages can we send, via our rootlets? What kind of helpful communication can we send out? We are living in a wildly interconnected world where, via technology, we can communicate with a wide variety of people quickly. Taking a lesson from trees, we can explore ways to connect naturally and to include and keep people and communities as part of the whole.

Seeding Change

Let's look at another example from nature—the sunflower. It starts off as a seed smaller than your fingernail and grows to a towering and marvelous flower, all in one growing season. It then creates multitudes of seeds to ensure more sunflowers the next year, and enough for many kinds of birds to enjoy. Sunflowers orient themselves to the sun throughout the day (thus their name), literally turning toward their source of energy.

What idea or source of inspiration could help our organization grow quickly and easily, like the sunflower? How can we create more opportunities (seeds) so that we have enough to share? How can our group turn toward more energy, whether that energy is more money, more people, a more involved community, or upgraded computers?

• • •

We saw new possibilities in the last few chapters for your green organization—new directions, new opportunities to bring people in. And now you have ideas and the beginnings of a plan for moving forward to invite more community and more diversity.

Congratulations!

Sprouts: Your Green Take-Home Tips

1. Answer the simple questions of who, what, when, where, why and how in order to create a plan to invite a new community to participate. Making a simple plan will help you see that it's possible to create a community-oriented, diverse, and lively organization.

2. Do some fun research—have coffee with someone in the new-to-you community, or go to an event that community is hosting, and find out more about them and their culture. Fun research and conversations yield fun and fruitful community events.

Next I'll share some tips on next steps to take and pitfalls to avoid in this adventure of protecting and honoring this amazing planet, and enjoying your journey on the way.

Chapter 8

Creating Community
Forward Steps

"We cannot change the world alone. To heal ourselves, to restore the earth to life, to create situations in which freedom can flourish, we must work together in groups."
–Starhawk

My wish for you is that you know you can create a community-centered, green, and thriving organization, and that you have taken powerful steps to start making that come true. I hope you have realized that people want to connect with your organization, and that you have the capacity to invite them, and you each have the capacity to contribute to the other.

May this book inspire you to take your first steps, or continue taking steps, along the journey to a diverse and community-oriented organization.

Renewing yourself is the first step. Then reinvigorating your relationships and, through those relationships, taking the steps to build, expand, and diversify your community.

We in the green movement stand for the community of diversity—the diversity of the exquisite and amazing animals and plants, birds and butterflies, with whom we share this planet. We will better protect those living jewels by honoring and listening to the wisdom of our own diversity, the diversity of humanity.

In the beautiful words of Alice Walker, (and her book of the same name), "Anything we love can be saved."

The Journey We've Taken Together

We've discovered, or perhaps rediscovered, how to cross barriers—through music, dance, words, and play.

We've practiced some simple but powerful thought-work tools to uncover and begin to loosen some of your own internal biases. Freeing your own mind allows you to think, do, and be different, more open, more willing to experiment with new ways to do things.

We have written new and empowering stories for ourselves and our organizations, stories we can live into and that encourage us to grow.

We've learned new ways to renew ourselves and our relationships, and we've explored how to build ties to new communities.

We've created plans to invite and engage a new community into our green organizations.

What's next? What can we do to ensure our success?

Step to Success

The first step can be summed up in one word: *Start*.

You and your green organization will not go anywhere different unless you take a first step. Yes, you can find out more information, learn more about the communities you want to connect with, but do something to begin reaching out, even as you do the research. If you research endlessly, without actually getting into community, that will only net you a lot of research, not necessarily a lot of connections or real community-building.

Stepping Forth Willingly

Be willing to make mistakes. Be a "fool willing." Do that crazy thang called "trial and error." Become familiar with missteps, mistakes, and "Oh well. We'll try again tomorrow." *That is the path.* There's more treasure shining along this fool-willing road than you'll be able to shake a stick at. You'll meet new friends and will establish wide and varied connections—possibly more than you've ever dreamed of. Your events will be talked of and anticipated. That is the lovely part, the part that will keep you going when things get challenging.

But you need to be willing to step in, and to change. You need to be willing to be uncomfortable. You need to be willing to step along the edges of what you think is possible. You need to reach out.

This is where it gets juicy, my friend. This is how the joy of life, of trying and failing and trying again, of starting something new, takes you home—home to yourself and home to a thriving community.

Community Building

You can bring diverse communities into your organization. You have a firm foundation for beginning. In the remainder of this last chapter, I'll invite you further, and show you some pitfalls to avoid in this journey.

Build Community in Community

This is not a do-it-yourself project. This is a do-it-together project. That may seem obvious. Or maybe not. We need to remember to renew ourselves, work on dissolving our internal biases, *and* build community while in community.

Build on a Slant

It's tempting to want to have a plan that's all perfect before putting it out into the world. I call this "the perfection misconception." We want to have it all be bright and shiny the first time through.

Let it be a bit misshapen. Let it be a bit slant. Do not let perfectionism stop you from starting your work with other communities. Do not let perfectionism keep you from playing, or from inviting others to play with you as you do your green work. Allow yourself to be the imperfectly perfect human that you are, and go from there. Strive for excellence, not perfection.

(*Excellence not perfection. Excellence not perfection.* That is something I chant to myself, as a practice. I need it, too.)

Do love what you are creating. It's much easier to put something out in the world when you love and are proud of something about it (even when it's not perfect), and when you are happy to have been part of creating it.

We've seen it happen when a team of people has done something wonderful. We see people beaming, or smiling with quiet pride. Some jump in the air to celebrate, others celebrate more quietly. People congratulate each other. There's a feeling of joy, of pride, a sense of celebrating a major accomplishment.

Perhaps you have seen this happen in events you've taken part in or hosted. Community members, friends, and families meet, some for the first time. People connect and talk about the event, tell silly stories, and laugh. It feels like everyone present had a hand in this circle of connection. The event of coming together for a common purpose creates new community, creates new bonds, and restores community.

Forget Isolation

You may try to make this change in isolation, or just by utilizing the people in your organization. Instead, I encourage you, once you have your bearings and have worked through some of the material in this book, to make this change alongside another organization, in partnership, *even while you are learning how to do it.*

You can take a stance that making change in community can be playful and inclusive. Have fun with this. Partner with another organization that is trying to do something similar or the same. Perhaps you have different modes of business, but you both want to bring more varied people and new ways of working together, to the table. *Perfect.* Put your heads together and brainstorm. Share resources, share work space, get together for creative problem-solving gatherings.

Partner with a group that has different strengths than yours does, so you can share, and help each other more, and grow together.

Partner with another human being who cares about this, so you can support each other, cheer each other on, and become one another's allies.

Build Representation

Remember why we are doing this. We want the voices, strengths, and talent of those who are underrepresented to be present in the process of healing the Earth, and participants in the joy of the process and in the results. We want them at the table. Or maybe we want to build a whole new table with them. Maybe including them will show us what that new table could look like and be like.

We want their talent, voices, and strengths to be part of our organizations. If we choose to invite more people to participate and to partner with us, that means we'll have more supporters, increased capacity, more financial support, increased innovation and creativity, and more allies and connections—we will be blessed in these gains. We can reflect, in our organizations and in the way we work together, our vibrantly multicultural world.

A World of Connections

In today's wildly interconnected world, there are so many ways to reach out to and be with other cultures and other people. There are conferences, dances, community potlucks, and multicultural celebrations. There are ways to connect through technology such as Skype, and Zoom, and FaceTime. And more ways to connect are being invented as you read this.

Here are some more ways to connect with others.

The Language of Action

Get into action. As we know, action speaks louder than words. Some communities have been studied, researched, questioned, and more… and then the researchers or organizers leave, and nothing that was needed by the community happened; nothing was improved. That defeats the purpose of finding out what's going on or what's going wrong. We want to actually change things for the better.

We can ask people in the communities we reach out to these important questions:

- What would you like to have change?
- Who would you love to have join you?
- What would you *love* to see go differently in your community?

Creating positive change in diverse communities will look different for each city, community, town, rural area, village, and group.

Woven Together

Imagine a single, suspended thread. Now imagine stepping on that thread. It would probably break, yeah? Now think of a beautiful woven rug made of many interwoven threads, and how strong and thick and beautiful it is. Imagine the vibrant colors, the rich patterns. Now imagine stepping on that rug, and how good it would feel on your feet, how welcoming.

Woven together, speaking up for each other in resilient community, we are stronger.

Building True Human Resources

We are building human resources when we reach out and work together—resources of cultural knowledge, wisdom, and willingness. We are building resiliency and strength, and the courage to change, to shift.

When you present yourself as a leader who is willing to learn, to change, to laugh, and to grow, you will magnetize people to you who are also willing—willing to speak up across borders, across barriers, and meet together to find ways to protect and restore this beautiful planet.

Tell yourself, *It's all good. I can start where I am. I can use my playful, creative self to encourage and create community connections. I can celebrate humor and invite different perspectives and contributions. I am part of a wide group of people around the world who are caring for and renewing the planet.*

People will respect you for being honest, for simply saying, "We are starting something new; we are doing things differently. We are excited, and we are learning as we do it. Please be patient with us as we learn and shift, as we grow. Thank you for joining us, and being part of this work and our community. We are delighted to have you here with us, and look forward to celebrating the accomplishments we create together.

Celebrations

I wish you all the best in playfully bringing the amazing diversity of human communities to your green organization, and for inventing new and creative ways to honor and restore this beautiful planet. I celebrate *you* for having the courage to reach out and make this world a better place. *Thank you* for coming along on this adventure with me.

Recommended Reading and Resources

"Play is our brain's favorite way of learning."
—Diane Ackerman

Books

- *Play: How it Shapes the Brain, Opens the Imagination, and Invigorates the Soul,* by Stuart Brown, MD, and Christopher Vaughan
- *Deep Play,* by Diane Ackerman
- *Finding Your Way in a Wild New World,* and *Finding Your Own North Star,* by Martha Beck

- *Truth or Dare: Encounters with Power, Authority, and Mystery*, by Starhawk
- *The Happiness Project*, by Gretchen Rubin
- *Loving What Is*, by Byron Katie
- *Anything We Love Can Be Saved*, by Alice Walker

Inspiring Organizations

- Badger Rock Community Center: www.resilientcities.org/projects-programs/badger-rock-center
- Urban Ecology Center: urbanecologycenter.org
- Sustain Dane: www.sustaindane.org

Thought Work Resource

- The Work by Byron Katie: thework.com/en

Resources on Race

- Center for Diversity and the Environment: www.cdeinspires.org
- Race Forward: www.raceforward.org

Acknowledgements

Thank you, with great appreciation and much love, to my parents, Pat and Gaylord Oppegard, for all of their love and support, and for encouraging me, by their example, to be a lifelong learner.

Thank you to my dear husband, David O'Donnell, whose love, support, and belief in me made this book possible. Thanks for caring for our kiddos so I could write!

Thank you to my children, Bjorn and Annika, for their joy, playfulness and vibrant enthusiasm. You two make my world go 'round. I love you! And for your big brother, Finnigan, our sweet boy already in the spirit world; this book is in honor of what would have been your tenth birthday on December 11, 2016. I love you Finn!

Thank you to Dr. Jane Goodall, for her inspiring work around the world with chimpanzees, and for making this world a better place for people. She is my hero.

I honor and celebrate Aung San Suu Kyi, for her immense persistence, bravery, and commitment to peace and democracy.

Thank you to Martha Beck, my mentor and teacher, for all of the laughter and wisdom she brings into the world. Thank you to the tribe of Martha Beck coaches from around the world, who bring laughter, love, warmth, and deep understanding to this human playground we are in. You're amazing and I love you folks! Special thanks to Master Coaches Anna Kunnecke, Pedro Baéz, Amy Pearson, Dr. Martha Jo Atkins, Indrani Goradia, Kris McGuffie, Dixie St. John, Abigail Steidley, Michael Trotta, Koelle Simpson, and Susan Hyatt for their kindness, inspiration, and wonderful work.

Thanks and great appreciation to the entire staff of Sustain Dane and Executive Director Jessie Lerner for their effective, interactive programs. Sustain Dane engages the community of Dane County, Wisconsin, to participate at their homes, workplaces and communities in making our world a better place for people and the planet. They are an amazing small team of people with a big, positive impact. (I am a huge fan!)

Thank you and kudos to Ken Leinbach, Director of the Urban Ecology Center, for walking his talk (quite literally, for he does not own a car). He and his team of dedicated staff and volunteers have created beautiful, welcoming places for both human and Earth diversity in Milwaukee, Wisconsin. The Center partners with schools to teach kids about their environment, celebrate citizen science (with bat, bird, and turtle research projects, to name a few), and lend outdoor equipment to those who want to try canoeing or ice skating, and much more.

Thank you to Marcia Caton Campbell and Hedi Rudd for welcoming me to the Badger Rock Neighborhood Center, of Madison, Wisconsin. The Center is a wonderful place of integral ecology, deep learning, and community. I loved watching the kids outside and enjoying the gardens and grounds; the Center has created a place for them, and they are becoming its stewards.

Thank you to Rinku Sen, Executive Director of Race Forward. I watched the video of her inspiring speech for the Bioneers Conference in 2015. She invited people to be more informed about race, just as she became more informed about the environment, and the intersections were insightful and amazing.

Thank you to the Bioneers, for their wildly creative, wonderful, and inspiring conference, speakers and materials.

Many thanks also to:

Joel Stone, my first supervisor at the Department of Natural Resources, who mentored me with kindness and pride.

My sixth grade teacher, Miss Karen J. Shellito, who believed in me.

My sister Laura, also known as LJ of O, for still trading favorite authors, books, and nicknames with me. I love you, Jelly Bean!

Sonya Newenhouse, who is an entrepreneur and sustainability expert, and founder of Wastecap Wisconsin, Community Car LLC, and the NewenHouse Kit Homes, for her inspiration and encouragement.

Thank you to dear friends Ingrid, Hannu & Kai Andersson, Sara Arscott, Susan Frikken, Krista Spiro, Jeanette Brynn, Kris McGuffie, Scott Caldwell, Monica Wightman, Elizabeth Doyle, Amy Gilliland, Stephen Montagna, Marion Farrior, and all of the dear ones in my life—you know who you are—who support and love me. (If I have missed someone, and I'm sure I have, please forgive the inadvertent omission. I still love 'ya.)

Some of my favorite wordsmiths—Ursula K. LeGuin, Maya Angelou, Sheri Tepper, J. K. Rowling, Rumi, Mary Oliver, James Herriot, Kenneth Grahame, and so many more, for giving me, with their generous creativity, countless hours of enjoyment, challenge, and inspiration.

The good people of American Players Theatre, who create wonderful theatre set on an outdoor, natural stage, near Springreen, Wisconsin, and who care for and restore the land around them.

Angela Lauria, my editor Grace Kerina, cover designer Jennifer Stimson, and the amazing people at the Difference Press, without whose encouragement, coaching, and belief, this book would not have been written. And a big shout out to my fellow writers and adventurers—your support and encouragement in this process have made it a worthy journey.

To the Morgan James Publishing team: Special thanks to David Hancock, CEO & Founder for believing in me and my message. To my Author Relations Manager, Margo Toulouse, thanks for making the process seamless and easy. Many more thanks to everyone else, but especially Jim Howard, Bethany Marshall, and Nickcole Watkins.

All of my beloved animal companions, especially Fudge the horse; dogs Clifford, Rascal, and Bucky; all of my sheep, especially Ribbon and Bowe; and kitties Tiny, Tiger, and Panda.

I thank and celebrate the magnificent burr oak tree on the campus of the University of Wisconsin at Madison, for its inspiration, and for the comfort it gave to this country girl when she came to the city to go to college.

And, finally, so much love and appreciation for the magnificent diversity of this exquisite planet—human, animal, plant, and the All of the Everything. I am delighted and honored to be a part of it all.

About the Author

Kathy Oppegard helps organizations within the environmental movement engage communities with their good work in the world, whether as volunteers, participants, or staff. She is a high-volume activist. Under her patient and enthusiastic management, literal tons of clothing have been collected for Goodwill Industries and tons of computer and electronics have been reused or recycled. She has coordinated hundreds of volunteers for America Recycles Day, and for Pollution Prevention Week for the state of Wisconsin, through the Department of Natural Resources.

Volunteers are some of Kathy's favorite people on this beautiful planet, and she has worked with over a thousand volunteers to help economically disadvantaged people get major repairs made on their home through Project Home.

Kathy is an Anglophile and book lover who has an extra scoop of wanderlust in her outdoorsy heart. She is a certified Martha Beck Life Coach, an entrepreneur, and a connoisseur of fair-trade dark chocolate.

Her friends describe her as delightful, thoughtful, and committed to positive change, and she is considered a go-to advisor for people who are ready to dream big dreams and create fun, practical advances for people and the planet.

Website: www.foolwillingbook.com
Email: coaching@foolwillingbook.com
Facebook: www.facebook.com/kathyoppegardcoaching

Thank You

Thank you so much for reading *Fool Willing: The Secret Power of Play to Engage Communities in Your Green Organization*. I hope the book has already inspired you.

I love it that you've arrived here. This tells me you're willing to create great solutions, have more fun, be more effective, and engage diverse communities in the restoration of this amazing planet. You're a member of what I like to call the Green Team.

To thank you, I've created the *Fool-Willing Toolkit*, a collection of free resources and ideas for practical fun to help you and your green organization engage in your community.

Get your free gifts and resources at *foolwillingbook.com/gifts*.

Go, Green Team!

Visit www.foolwillingbook.com/gifts
for free resources in the *Fool-Willing Toolkit*
to help you build and celebrate your community.

Morgan James
Speakers Group

↗ www.TheMorganJamesSpeakersGroup.com

We connect Morgan James published authors with live and online events and audiences whom will benefit from their expertise.

Morgan James makes all of our titles available
through the Library for All Charity Organization.

www.LibraryForAll.org

Printed in the USA
CPSIA information can be obtained
at www.ICGtesting.com
JSHW082349140824
68134JS00020B/1976